T0331028

Introduction to Classifier Performance Analysis with R

Classification problems are common in business, medicine, science, engineering, and other sectors of the economy. Data scientists and machine learning professionals solve these problems through the use of classifiers. Choosing one of these data-driven classification algorithms for a given problem is a challenging task. An important aspect involved in this task is classifier performance analysis (CPA).

Introduction to Classifier Performance Analysis with R provides an introductory account of commonly used CPA techniques for binary and multiclass problems, and use of the **R** software system to accomplish the analysis. Coverage draws on the extensive literature available on the subject, including descriptive and inferential approaches to CPA. Exercises are included at the end of each chapter to reinforce learning.

Key Features:

- An introduction to binary and multiclass classification problems is provided, including some classifiers based on statistical, machine, and ensemble learning.
- Commonly used techniques for binary and multiclass CPA are covered, some from less well-known but useful points of view. Coverage also includes important topics that have not received much attention in textbook accounts of CPA.
- Limitations of some commonly used performance measures are highlighted.
- Coverage includes performance parameters and inferential techniques for them.
- Also covered are techniques for comparative analysis of competing classifiers.
- A key contribution involves the use of key **R** meta-packages like **tidyverse** and **tidymodels** for CPA, particularly the very useful **yardstick** package.

This is a useful resource for upper-level undergraduate and masters level students in data science, machine learning, and related disciplines. Practitioners interested in learning how to use **R** to evaluate classifier performance can also potentially benefit from the book. The material and references in the book can also serve the needs of researchers in CPA.

CHAPMAN & HALL/CRC DATA SCIENCE SERIES

Reflecting the interdisciplinary nature of the field, this book series brings together researchers, practitioners, and instructors from statistics, computer science, machine learning, and analytics. The series will publish cutting-edge research, industry applications, and textbooks in data science.

The inclusion of concrete examples, applications, and methods is highly encouraged. The scope of the series includes titles in the areas of machine learning, pattern recognition, predictive analytics, business analytics, Big Data, visualization, programming, software, learning analytics, data wrangling, interactive graphics, and reproducible research.

Recently Published Titles

Introduction to NFL Analytics with R
Bradley J. Congelio

Soccer Analytics
An Introduction Using R
Clive Beggs

Spatial Statistics for Data Science
Theory and Practice with R
Paula Moraga

Research Software Engineering
A Guide to the Open Source Ecosystem
Matthias Bannert

The Data Preparation Journey
Finding Your Way With R
Martin Hugh Monkman

Getting (more out of) Graphics
Practice and Principles of Data Visualisation
Antony Unwin

Introduction to Data Science
Data Wrangling and Visualization with R Second Edition
Rafael A. Irizarry

Data Science
A First Introduction with Python
Tiffany Timbers, Trevor Campbell, Melissa Lee, Joel Ostblom and Lindsey Heagy

Mathematical Engineering of Deep Learning
Benoit Liquet, Sarat Moka, and Yoni Nazarathy

Introduction to Classifier Performance Analysis with R
Sutaip L.C. Saw

For more information about this series, please visit: https://www.routledge.com/Chapman--HallCRC-Data-Science-Series/book-series/CHDSS

Introduction to Classifier Performance Analysis with R

Sutaip L.C. Saw

CRC Press
Taylor & Francis Group
Boca Raton London New York

CRC Press is an imprint of the
Taylor & Francis Group, an **informa** business
A CHAPMAN & HALL BOOK

First edition published 2025
by CRC Press
2385 Executive Center Drive, Suite 320, Boca Raton, FL 33431

and by CRC Press
4 Park Square, Milton Park, Abingdon, Oxon, OX14 4RN

CRC Press is an imprint of Taylor & Francis Group, LLC

© 2025 Sutaip L.C. Saw

Reasonable efforts have been made to publish reliable data and information, but the author and publisher cannot assume responsibility for the validity of all materials or the consequences of their use. The authors and publishers have attempted to trace the copyright holders of all material reproduced in this publication and apologize to copyright holders if permission to publish in this form has not been obtained. If any copyright material has not been acknowledged please write and let us know so we may rectify in any future reprint.

Except as permitted under U.S. Copyright Law, no part of this book may be reprinted, reproduced, transmitted, or utilized in any form by any electronic, mechanical, or other means, now known or hereafter invented, including photocopying, microfilming, and recording, or in any information storage or retrieval system, without written permission from the publishers.

For permission to photocopy or use material electronically from this work, access www.copyright.com or contact the Copyright Clearance Center, Inc. (CCC), 222 Rosewood Drive, Danvers, MA 01923, 978-750-8400. For works that are not available on CCC please contact mpkbookspermissions@tandf.co.uk

Trademark notice: Product or corporate names may be trademarks or registered trademarks and are used only for identification and explanation without intent to infringe.

ISBN: 978-1-032-85562-2 (hbk)
ISBN: 978-1-032-85010-8 (pbk)
ISBN: 978-1-003-51867-9 (ebk)

DOI: 10.1201/9781003518679

Typeset in CMR10 font
by KnowledgeWorks Global Ltd.

For my wife, Cecilia.
Thank you for your support and help with this book.

Contents

List of Tables

List of Figures

Preface

Welcome to *Introduction to Classifier Performance Analysis with* **R**. This book is for those who want a reasonably complete (at least at an introductory level) and up-to-date coverage on the analysis of classification algorithms through the use of performance measures and curves. It attempts to synthesize useful material from the vast published literature on the subject. Another motivation for the book is to show how **R** can be used to perform the required analysis. As computational software, **R** has already demonstrated its excellence to a large international community of users. Its appeal is further enhanced by recently developed packages and meta-packages for data science, machine learning, and classification performance analysis in particular.

Relevance & Importance of the Topic

Given the recent advances in computing technology and power, disciplines like data science (DS) and machine learning (ML) have grown to be increasingly relevant, important, and useful. Demand for professionals in these disciplines is high as evidenced by information in the following web links:

- The Data Scientist Job Market in 2024
  https://365datascience.com/career-advice/data-scientist-job-market/

- Companies Are Desperate for Machine Learning Engineers
  https://builtin.com/data-science/demand-for-machine-learning-engineers

Classification problems are common in applications like fraud detection, target marketing, credit scoring, disease detection, customer churn prediction, spam filtering, and quality control (this list is not exhaustive). Data scientists and machine learning professionals solve these problems through the use of classifiers, i.e., data-driven classification models or algorithms that facilitate prediction of class labels and membership probabilities based on the features of cases (e.g., individuals or objects).

Choosing the right classifier for a given problem is a challenging task. A critical aspect of what is involved is classifier performance analysis (CPA). Such analysis employs performance measures/curves and resampling techniques to determine if a trained classifier is doing a good enough job at classifying unseen cases. These techniques are also relevant when a comparative analysis is made of competing classifiers, and when new classification algorithms are evaluated.

This book provides an introduction to CPA and the use of the **R** software system to accomplish the analysis. Much of what is known about CPA is scattered throughout the published literature, including books and journals in disciplines other than DS and ML. To have the relevant material in one book, with expanded introductory discussions and elementary theoretical support where necessary, and illustrated with help from **R** is certainly useful for those who have to engage in CPA, particularly those who have yet to master the techniques and conceptual foundations underlying the analysis.

Target Audience

The book is primarily targeted at senior undergraduate and masters level students in DS and/or ML (it is not a monograph on the CPA for professionals in these and related disciplines). It can serve as a supplementary text or reference, especially since coverage of CPA is somewhat limited in most published introductory textbooks on DS and ML.

Aspiring data scientists, machine learning professionals, and others who have to analyze performance of classification algorithms should find the book a useful resource. Early career professionals in DS, ML, and related disciplines will find material in the book that can help consolidate their understanding of CPA. Practitioners and researchers, especially the experienced ones, will probably be familiar with much if not all of the material in this introductory book. However, they can also potentially benefit from the book, e.g., using it as a reference for the use of key packages and meta-packages in **R** for CPA (the extensive bibliography given for this topic is also useful).

Why R?

Python is a programming language that is commonly used by DS and ML professionals, and it does a great job for what it was designed for based on publications and information in alternative media the author has seen on this software. However, **R** and its superlative integrated development environment **RStudio** offers some appealing competitive advantages. Its excellence in serving the needs of analysts in DS and ML and other disciplines engaged in computational statistics and data mining is unquestionable. In particular, it can be used to train a wide variety of classifiers and, for our purpose, it provides a powerful and well-integrated collection of tools for binary and multiclass CPA given the availability of packages like **yardstick** and meta-packages like **tidyverse** and **tidymodels**.

On a personal note, the author regards **R** the best choice for students learning to solve problems in computational statistics, data science, machine learning, and related disciplines. When making this judgement, he has other software systems (like **Fortran**, **APL**, **C**, **Visual Basic**, **Gauss**, **SAS**, **SPSS**, and **Stata**) in mind that he has used in an educational environment. This judgement remains unchanged when his experience with **S** and **S-Plus** is also taken into account, even though (like **S-Plus**) **R** is based on **S**.

Chapter Outlines

- Chapter 1 provides a brief introduction to classification problems and classifiers, and gives an overview of classification performance analysis (CPA) for binary and multiclass problems.

- Chapter 2 focuses on performance measures and parameters for binary CPA. Descriptive and inferential techniques are demonstrated using functions in the **yardstick** package. This **R** package and meta-packages like **tidyverse** and **tidymodels** are also used in subsequent chapters.

- Chapter 3 covers some commonly used performance curves in binary CPA, including established and more recent performance measures derived from them.

- Chapter 4 discusses the comparative analysis of four classifiers (comprising a mixture of statistical, machine, and ensemble learners) for a binary classification problem.

- Chapter 5 covers descriptive and inferential techniques for analyzing performance of multiclass classifiers. The coverage includes performance measures and curves that apply to the multiclass problem.

- Chapter 6 ties up some loose ends with discussions on modeling issues, resampling techniques, and class imbalance (including other topics that deal with the wider issue of classifier performance like cost-sensitive learning).

What can Readers Learn?

- What can students in DS, ML, and related disciplines hope to learn from the book?

  - The relevance and importance of classification problems and how to solve them with data-driven classifiers.

  - The different categories of classification algorithms and examples of some classifiers that are based on them.

  - The key measures/curves and resampling techniques that can be used to assess generalizability of trained binary and multiclass classifiers.

  - Relative merits of various performance metrics (i.e., measures and curves) for CPA and the impact of class imbalance on the metrics.

  - Alternative and complementary views on the measures and curves that CPA relies on.

  - Some techniques for statistical inferences on performance parameters.

  - Key modeling issues that impact on training and evaluation of classifiers.

  ○ How to use state-of-art functionality in **R** to train classifiers, perform CPA and compare classifiers, in particular, use of the **yardstick** package and meta-packages like **tidyverse** and **tidymodels**.

- What can others hope to benefit from the book?

  ○ Newly trained DS/ML practitioners whose training was based on **Python** may wish to learn **R** to expand their skillset, and using **R** for CPA (and classifier training) is one of several ways to achieve this goal. Furthermore, gaps in their knowledge of CPA (which possibly exists, given the limited scope of material on CPA in textbook accounts) can be filled by the book.

  ○ The book can be useful for instructors interested in an expanded or complementary coverage of CPA in courses they teach; hence, the utility of this book as supplementary text. Furthermore, the exercises provided in the book can be used for student assignments.

  ○ Experienced practitioners interested in learning **R** to train and evaluate classifiers can use the book as an introduction to the extensive capabilities of **R** to solve these (and other) problems.

  ○ Users of classifiers need to be reasonably familiar with CPA in order to question and understand performance of proposed classification algorithms for their problems. The book can provide the required background information to help evaluate proposed classifiers.

Supplementary Materials

Readers can access soft copies of code in the book and those for exercises, including relevant datasets from the publisher's website.

Acknowledgments

I am grateful to commissioning editor Lucy McClune and editorial assistant Danielle Zarfati from Taylor and Francis for the significant roles they played with publication aspects related to this book. Without Lucy's expeditious response to my book proposal and her helpful guidance in getting the book project underway, this book would probably not see the light of day. Danielle was of great help in this endeavor with her prompt and helpful responses to the numerous questions I had regarding publication issues. I also wish to thank Ashraf Reza and the production team for not only the good job they did in preparing the proofs for the book, but also for their help in dealing with the issues that arose from the proofs.

Author

Sutaip L. C. Saw holds a PhD from The Wharton School, University of Pennsylvania. Prior to earning his PhD, he served as a statistician in the public sector. His subsequent career was spent as an academic with research interests and publications in engineering statistics and statistical computing, and he has significant teaching experience in statistical/mathematical subjects at undergraduate and postgraduate levels.

Since leaving academia, he has been focused on applications of **R** to data mining and machine learning problems. Although his interest in classification problems and performance analysis of classifiers started while he was still an academic, it has intensified in recent years and this book is the result of time spent on the topic at hand.

1

Introduction to Classification

Assigning an individual or object to one of several classes is a task that occurs frequently in situations where an analyst is focused on problems like fraud detection, target marketing, credit scoring, disease detection, customer churn prediction, spam filtering or quality control. You have a classification problem when you are faced with any one of these important practical problems.

When faced with one of the problems mentioned above, how does one solve it? Given sufficient information, you can use a data-driven classifier to solve the classification problem. An important issue that arises when you take such an approach is the question of classifier performance. How can you determine whether the classifier you have for the problem is doing a good enough job at classifying unseen cases? In this book, we discuss techniques in classifier performance analysis (CPA) that you can use to answer a question like this.

1.1 Classification Overview

You can use a statistical or machine learning approach to solve a classification problem. In practice, several techniques from these and other approaches are tried and the best-performing classifier is selected for the problem. Regardless of the approach taken, you need training data to construct the classifier and separate test data to evaluate performance of the trained classifier. These datasets contain information on the features (i.e., predictors) and target variable (or class label) of cases that you want to classify.

The binary classification problem is a basic one since it facilitates development of many ideas and techniques regarding classifier construction and evaluation. A good understanding of what is involved in this problem is needed before taking up similar issues for the multiclass problem. We provide an overview of both types of problems in the next and subsequent sections of this chapter. Except for Chapter 5, all remaining chapters will focus on the 2-class problem.

DOI: 10.1201/9781003518679-1

1.1.1   Binary and Multiclass Problems

You have a binary classification problem when you have to assign a case to one of two classes based on information from a set of features for the case. Such problems arise when you face questions like those listed below (the accompanying references contain relevant examples) in a practical situation.

- Does a particular individual have coronary heart disease? [1, 80]

- Will an applicant for credit default on repayment? [55]

- Does a pair of database records match (i.e., refer to the same real-world entity)? [46]

- Will a randomly selected passenger on a cruise liner survive if it sinks? [95, p. 98]

- Will an individual default on his/her credit card payment? [62, p. 133]

- Is a particular e-mail spam? [119, p. 175]

When faced with a binary classification problem, you can begin by identifying the class of interest, and when presented with a case, ask the question whether the case belongs to the class of interest.[1] You can regard a "Yes" response to mean that the case belongs to the *positive* class regardless of the substantive nature of the class in question. For example, the two classes may be labeled as "spam" and "non-spam" for a spam filtering problem. If the former is the class of interest, then you can label it as the *positive* class even though the usual thinking is to regard getting spam as a negative outcome.

There are, however, alternative criteria that have been used to define the *positive* class. For imbalanced binary problems, this class is usually the minority class since it is often the one of primary interest; for example, in medical diagnosis, the focus is usually on an individual belonging to a (minority) diseased group. This convention is widely adopted by data scientists and machine learning professionals. The sidebar titled *Bad Positives and Harmless Negatives* in Provost and Fawcett [90, p. 188] provides an informative perspective on this issue. Whatever convention is adopted, it is important to make it clear when using class-specific performance measures to evaluate binary classifiers because such measures attempt to quantify the rate at which true positives or true negatives occur, for example.

In practice, you can also encounter problems that involve a target variable with more than two levels. Such multiclass classification problems can arise when you have questions like the following ones.

---

[1] When discussing classification problems in this book, a case refers to an individual or object, or some other entity. Readers in epidemiology, for example, should be mindful since the term has a different meaning in their discipline. A case is often referred to as an instance when data mining is used to solve classification problems. It has also been referred to as an example in other application areas.

- Is an individual normal, chemically diabetic, or overtly diabetic? [95, p. 60]

- Is the income level of an individual low, medium, or high? [83, p. 206]

- Is the labor force participation status of an individual unemployed, working part time, or working full time? [41, p. 260]

1.1.2 Classification Rules

Decision scientists and machine learning professionals use classifiers to solve classification problems that arise in practice. You can think of classifiers as data-driven classification rules derived from supervised learning techniques; see [42, 47, 113] for further discussion on such techniques.

There are a variety of classifiers that you can choose from. Some of these will be discussed in this and subsequent chapters. However, our interest in this book is not in a detailed discussion of the various classification techniques since adequate coverage may be found, for example, in books on statistical [59, 62] and machine [6, 83, 95] learning. Rather, our focus will be on the analysis of classifier performance. Further discussion of what is involved requires introduction of some notation and terminology.

To begin, let Y denote the unknown target (response) variable for a randomly selected case with feature vector $\boldsymbol{X}$, and let $\widehat{Y}$ denote the corresponding classifier predicted target (whose value determines the class for the case). The predicted response is a random variable defined by

$$\widehat{Y} = \hat{f}(\boldsymbol{X}), \tag{1.1}$$

where $\hat{f}(\cdot)$ is a classifier prediction function that is determined by a suitable dataset. We assume that Y and $\widehat{Y}$ can take values in $\mathcal{C}$ where $\mathcal{C} = \{1, 2\}$ for binary classification, and $\mathcal{C} = \{1, \ldots, k\}$ with $k > 2$ for the multiclass problem.[2] For convenience, we will regard the first element in $\mathcal{C}$ as the reference class (this is the *positive* class for binary classification). When working with target variables, we are essentially treating our classification problem as one concerned with prediction of a nominally valued variable. This is so despite the fact that it is not uncommon to work with a labeled class variable when solving practical classification problems.

In practice, binary classification problems are often solved using scoring classifiers that make class assignments based on classification rules of the form:

Assign a case with feature vector $\boldsymbol{x}$ to the *positive* class if $S(\boldsymbol{x}) > t$,

---

[2]For binary classifiction, other choices for $\mathcal{C}$ that have been adopted in the literature include $\{0, 1\}$ and $\{-1, 1\}$. The former choice, though convenient when modeling a Bernoulli target variable, is not mandatory. Our choice is consistent with what is given here for the multiclass problem.

where $S(\boldsymbol{x})$ is a score assigned by the classifier to a case with given feature vector $\boldsymbol{x}$ (this vector contains attributes of a case that are relevant for the classification problem) and t is a suitable threshold. The prediction function associated with the above classification rule may be expressed as

$$\hat{f}(\boldsymbol{x}) = 2 - I\left(S(\boldsymbol{x}) > t\right), \tag{1.2}$$

where the zero-one indicator function $I(\cdot)$ takes the value one if its argument is true. As seen by (1.2), the score given by the classifier to a case is instrumental in determining class assignment. A score may be the estimated *positive* class membership probability or a suitable scalar quantity that measures the strength of *positive* class membership for a case. Examples will be given later in our discussion of the logit model classifier.

As demonstrated in the next section, we use training data to determine the score function (depending on the classifier, this empirically determined function may be some thing simple or complicated and nonlinear). When training statistical learners, you typically focus effort on estimating parameters of the underlying model by optimizing a suitable statistical criterion (e.g., likelihood function); for other classifiers (e.g., those based on machine learning), your goal is to obtain the "best" settings for the hyperparameters that drive the underlying classification algorithm.

The training dataset refers to a set $\{(\boldsymbol{x}_1, y_1), \ldots, (\boldsymbol{x}_n, y_n)\}$ from n cases where $\boldsymbol{x}_i$ is the *observed* feature vector for the i-th case and y_i is the corresponding *observed* value of the target variable. See Figures 1.1 and 1.2 for simulated training data for a synthetic binary and 3-class problem, respectively. For both problems, the feature vector is rather simple and may be expressed as $(age, income)$, and $group$ (with number of classes depending on the problem) is the class variable. For both problems, $group$ is a categorical representation of the target variable (the reference class is "Yes" for the binary problem, and "A" for the 3-class problem). Notice that you need information on the features and class (or target) variables of cases when you train a classifier. That is why such training is called supervised learning.[3] Once a classifier is trained (and tested), it can be subsequently used to classify cases with *unknown* target values.

Finally, note that it is customary to evaluate a classifier after training it. This involves the use of the trained classifier and test data to obtain performance measures and curves in a process called classifier performance analysis (CPA). In this chapter, we give an overview of what is involved in CPA. Our focus will be on various ideas underlying the analysis without worrying (for pedagogical reasons) about how to obtain the presented results. Subsequent chapters will provide further details on the numerical and graphical techniques used in CPA including **R** code to obtain the various results.

---

[3]In unsupervised learning, like what is involved when you solve a clustering problem, you only have information on the features of cases.

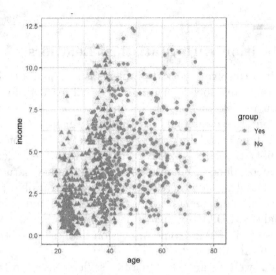

FIGURE 1.1
Training Data for a Binary Classification Problem

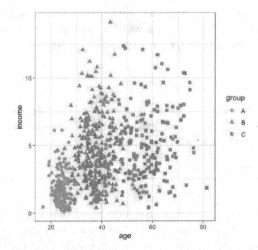

FIGURE 1.2
Training Data for a 3-Class Classification Problem

1.2 Classification with Statistical Learners

Learning from data to solve practical problems is a common task in business, science, engineering, and social science. Statistical learners for classification problems rely on models and principles from statistics. For example, logit model classifiers are based on logistic regression, a member of the GLM family of generalized linear models. Naive Bayes classifiers rely on an important result in probability theory called Bayes theorem with the added assumption of conditional independence. Dimension reduction is a key data analytic principle in statistical analysis and there are several ways to apply it. Discriminant Analysis (DA) is one implementation of this principle, and it leads to linear discriminant analysis (LDA) and quadratic discriminant analysis (QDA) classifiers; see James et al. [62] for some discussion of these and other statistical learners for classification.

In this section, we give an overview of classifiers based on logistic regression for binary and multiclass problems. We focus on this category of classifiers for several reasons. Exposition is simpler compared to what is involved with, for example, LDA classifiers, i.e., those based on Fisher's linear discriminant analysis [39, 48]. Some early attempts at solving classification problems were based on LM classifiers, and so like LDA classifiers, they represent a noteworthy category of classification techniques. Readers with some exposure to statistics at the intermediate level (a level we assume them to have) should be able to understand the discussion of these classifiers without too much difficulty. Admittedly, in practice, they may not be as powerful as some other classifiers that we will encounter in later chapters, but note that simple classifiers should not be readily dismissed since they can perform as well as more sophisticated ones [49]. Thus, it is a worthwhile effort for those interested in classification techniques to start with statistical learners.

1.2.1 Logit Model Classifier

As suggested by its name, the logit model (LM) classifier is based on a logit model (a special case of logistic regression). For binary classification, we model the logit of *positive* class membership for a case with feature vector $\boldsymbol{x}$ by

$$\ln \frac{P(Y=1)}{P(Y=2)} = \eta(\boldsymbol{x}), \qquad (1.3)$$

where $\eta(\boldsymbol{x})$ represents the linear predictor and Y is a target variable that takes the value 1 or 2 according to whether the case belongs to the *positive* or *negative* class, respectively. The linear predictor determines the logit model, and it is a key specification that is needed for the corresponding classifier. It is determined by the features of a case and the parameters of the model; see (1.5) for a simple example.

It follows from (1.3) that the probability that a case belongs to the *positive* class is

$$P(Y = 1) = \frac{\exp(\eta(\boldsymbol{x}))}{1 + \exp(\eta(\boldsymbol{x}))}.$$

Thus, a case may be assigned to the *positive* class if this probability is sufficiently large. This amounts to saying that such a class assignment is made by a LM classifier if

$$\frac{\exp(\eta(\boldsymbol{x}))}{1 + \exp(\eta(\boldsymbol{x}))} > c, \tag{1.4}$$

where c is a suitable threshold. The usual default for c is 0.5, but note that this value is not suitable when you have significant class imbalance in your problem. During our discussion on cost-sensitive learning in the last chapter, we show one way to obtain an optimal value for this threshold.

To illustrate training of a logit model (LM) classifier, consider the data that is displayed in Figure 1.1. A partial listing of the data is given below.

```
# Training Data for a Binary Classification Problem

## # A tibble: 700 x 3
##      age income group
##    <dbl>  <dbl> <fct>
## 1  23.3   3.60 No
## 2  39.5   4.24 Yes
## 3  22.9   3.00 No
## # ... with 697 more rows
```

For the binary classification problem under consideration, the feature vector is $\boldsymbol{x} = (age, income)$, and a possible linear predictor (we ignore possible interaction effects) is

$$\eta(\boldsymbol{x}) = \beta_0 + \beta_1 \times age + \beta_2 \times income. \tag{1.5}$$

The β_i's on the right-hand side of the above expression are unknown model parameters. Here, we use the training data and the `glm()` function in the **stats** package to obtain estimates of these parameters; a good discussion on estimation of logit model parameters may be found in Charpentier and Tufféry [12, p. 175], for example.[4]

When we use the training data to estimate the parameters in linear predictor (1.5), we obtain

$$\hat{\beta}_0 = -11.4, \hat{\beta}_1 = 0.301 \text{ and } \hat{\beta}_2 = -0.024.$$

---

[4]When using the `glm()` function in the **stats** package to fit the logit model, keep in mind that the function models the logit of the second level of the categorical class variable and make any required adjustments, i.e., let the second level of the class variable refer to the *positive* class and change later when using **yardstick** package to evaluate performance, if required (the reason for the change will be explained later).

We can assess the statistical significance of the estimates by looking at the corresponding P-values.

```
# Logit Model Parameter Estimation

## # A tibble: 3 x 3
##    term        estimate  p.value
##    <chr>          <dbl>    <dbl>
## 1 (Intercept)   -11.4   2.48e-32
## 2 age             0.301  1.44e-31
## 3 income         -0.0240 6.29e- 1
```

On examining the P-values, we note that the estimated coefficient for *income* is not statistically significant. Refitting the model without this feature yields the following results (notice that the estimated *intercept* and *age* coefficients change only slightly when *income* is omitted).

```
# Logit Model Parameter Estimation (cont'd)

## # A tibble: 2 x 3
##    term        estimate  p.value
##    <chr>          <dbl>    <dbl>
## 1 (Intercept)   -11.5   2.88e-32
## 2 age             0.299 2.15e-32
```

In light of the above results, we see that the trained LM classifier is one which assigns an individual to the *positive* class if

$$\frac{\exp\{-11.5 + 0.299 \times age\}}{1 + \exp\{-11.5 + 0.299 \times age\}} > c. \tag{1.6}$$

The left-hand side of (1.6) represents the estimated *positive* class membership probability of an individual with a given value for *age*. We can re-express (1.6) as

$$-11.5 + 0.299 \times age > t, \tag{1.7}$$

where $t = \ln(c/(1-c))$. Thus, the LM classifier is a scoring classifier with scores that can be defined by the left-hand side of (1.6) or (1.7). When scores are probabilities, the classification rule is also referred to as a probabilistic classifier [16]. The trained LM classifier given by (1.6) with $c = 0.5$ will be used to illustrate some techniques for evaluating classifier performance in later sections of this chapter.

Before considering another example of classifier training in the next section, it should be noted that such training in practice is often an involved process that requires you to do more than just apply a classification algorithm to a training dataset. This is certainly true if you want to obtain the best performing classifier for your problem. Often, if that is your practical objective, you will also need to do some combination of the following: feature engineering, data exploration and preprocessing, hyperparameter tuning (this requires use of resampling methods like bootstrapping or cross-validation), and selection of suitable performance measures to evaluate the trained classifier.

To avoid unnecessary complications in this book, we will take a limited approach to train the classifiers that will be used to illustrate various CPA techniques (when applicable, we will discuss some of the abovementioned training aspects in subsequent chapters). Clearly, this is not unreasonable since it suffices to have a trained classifier for such illustrations (i.e., you do not need to have the "best" classifier).

1.2.2 Multinomial Logistic Classifier

The multinomial logistic (ML) classifier for a k-class classification problem assigns a case to the h-th class if

$$h = \operatorname*{argmax}_{i} \{P(Y = i \mid \boldsymbol{x}), \ i = 1, \ldots, k\}$$

where

$$P(Y = i \mid \boldsymbol{x}) = \frac{\exp(\eta_i(\boldsymbol{x}))}{\sum_{j=1}^{k} \exp(\eta_j(\boldsymbol{x}))}, \quad i = 1, 2, \ldots, k.$$

Here, $\boldsymbol{x}$ is the feature vector for a case, $\eta_1(\boldsymbol{x}) = 0$, and

$$\eta_i(\boldsymbol{x}) = \beta_{0i} + \beta_{1i}x_1 + \cdots + \beta_{mi}x_m, \quad i = 2, 3, \ldots, k,$$

are linear predictors such that

$$\ln \frac{P(Y = i \mid \boldsymbol{x})}{P(Y = 1 \mid \boldsymbol{x})} = \eta_i(\boldsymbol{x}).$$

Further information on this statistical learner may be found in Fox and Weisberg [41, p. 260], for example. For applications of this classification learner, see [31, 62, 65].

To illustrate training of an ML classifier, consider the data that is displayed in Figure 1.2. A partial listing of the data is given below.

```
# Training Data for a 3-Class Classification Problem

## # A tibble: 700 x 3
##      age income group
##    <dbl>  <dbl> <fct>
## 1   23.9  0.307 A
## 2   27.2  0.390 A
## 3   24.1  0.671 A
## # ... with 697 more rows
```

The following estimated parameters of the multinomial logistic regression model were obtained when the model was fitted to the given training data.

```
# Multinomial Logistic Regression Parameter Estimation

## # A tibble: 6 x 4
##    y.level term           estimate  p.value
##    <chr>   <chr>             <dbl>    <dbl>
## 1 B       (Intercept)      -45.6  2.41e- 6
## 2 B       age                1.51 4.15e- 6
## 3 B       income             1.10 3.31e- 3
## 4 C       (Intercept)      -71.6  1.44e-12
## 5 C       age                2.05 8.58e-10
## 6 C       income             1.14 3.07e- 3
```

These results show that the estimated model coefficients are statistically significant. Thus, the estimated probabilities of class membership are given by (class "A", "B" and "C" are referred to as class 1, 2 and 3, respectively)

$$\frac{\exp(\hat{\eta}_i(\boldsymbol{x}))}{\sum_{j=1}^{3} \exp(\hat{\eta}_j(\boldsymbol{x}))}, \quad i = 1, 2, 3,$$

where $\hat{\eta}_1(\boldsymbol{x}) = 0$,

$$\hat{\eta}_2(\boldsymbol{x}) = -45.6 + 1.51 \times age + 1.10 \times income,$$

and

$$\hat{\eta}_3(\boldsymbol{x}) = -71.6 + 2.05 \times age + 1.14 \times income.$$

Hence, the trained ML classifier assigns a case to the h-th class if

$$h = \underset{i}{\operatorname{argmax}} \left\{ \frac{\exp(\hat{\eta}_i(\boldsymbol{x}))}{\sum_{j=1}^{3} \exp(\hat{\eta}_j(\boldsymbol{x}))}, \ i = 1, 2, 3 \right\}. \tag{1.8}$$

We will use the above classifier to provide an overview of multiclass CPA in Section 1.4.

1.3 An Overview of Binary CPA

The evaluations involved in binary classifier performance analysis are motivated by questions like those given below (as noted earlier, *positive* cases belong to the class of interest for a substantive problem).

- What fraction of cases in an evaluation dataset are correctly classified?

- What fraction of *positive* cases are correctly classified?

- What fraction of *positive* classifications are correct?

- What is the degree of concordance between actual and predicted classes?

The above list of questions is clearly not exhaustive; others will be mentioned in due course. In this section, we show how to answer the first two questions through the use of descriptive classifier performance measures like *accuracy* and *sensitivity*. The measures that you can use to answer the remaining two questions and others will be discussed later in the next chapter.

The measures and techniques for CPA are quite well established in data science and machine learning. For a good introductory account of these topics, see Chapter 9 in [83]. Useful additional information may be found in published review articles like those in [16, 52, 63, 74, 102, 103].

1.3.1 Required Information

The information you need for CPA (whether binary or multiclass) may be obtained in a number of ways. We begin by considering the validation set (VS) approach; you can also use resampling techniques to obtain the required information. In the VS approach, the available data is randomly divided into a training set and a validation or hold-out set [62]. A classifier is trained using the first dataset, and the trained classifier is evaluated using the second dataset (the latter is also referred to as a test or evaluation set). As is well known among data scientists and machine learning professionals, information in the test set should not be used during classifier training in order to avoid unnecessary bias when the trained classifier is evaluated.

To evaluate a trained binary classifier, the minimum information you need are the actual and predicted classes for cases in a test dataset. For a more complete evaluation, you also need to have predicted *positive* class membership probabilities (or other suitable class membership scores).

For example, consider the trained LM classifier given by (1.6). To assess its generalizability (i.e., ability to classify new cases) of the classifier, you can use information in the (partially listed) 300×3 tibble given below.

```
# Required Information for Binary CPA

## # A tibble: 300 x 3
##    prob_Yes pred_class group
##       <dbl> <fct>      <fct>
## 1     0.143 No         Yes
## 2     0.531 Yes        Yes
## 3     0.139 No         Yes
## # ... with 297 more rows
```

The given tibble was obtained when the LM classifier was applied to the cases in a test dataset. The column labeled `prob_Yes` in the partially listed tibble contains predicted *positive* class membership probabilities for the 300 cases in the test dataset, and the second column labeled `pred_class` contains the corresponding predicted classes. Information on the actual classes are contained in the third column. With the given information, you can obtain a useful array summary and extract descriptive performance measures from it. You can also construct suitable performance curves and obtain relevant summaries from them.

1.3.2   Binary Confusion Matrix

The confusion matrix is an important array summary that plays an important role in the performance analysis of a trained classifier. It is derived from data on predicted and actual classes of cases in an evaluation dataset (e.g., test dataset). A generic representation of this summary for binary classifiers is given in Table 1.1. This table is what statisticians refer to as a cross-tabulation of predicted versus actual classes. For those interested in CPA, the table entries labeled tp, fn, fp, and tn represent counts that make up the 2×2 array that is usually referred to as a confusion matrix. You can derive useful descriptive summaries from this matrix to help you evaluate performance of a trained classifier.

You can interpret the four key counts in a binary confusion matrix as follows:

$$
\begin{aligned}
tp &= \text{number of true positives,} \\
fn &= \text{number of false negatives,} \\
fp &= \text{number of false positives,} \\
tn &= \text{number of true negatives.}
\end{aligned}
$$

True positives (negatives) refer to those actual *positive* (*negative*) cases that were correctly classified. It follows false positives (negatives) refer to those actual *negative* (*positive*) cases that were incorrectly classified. Clearly, false positives and negatives refer to errors that you can make in binary classification problems. Related terminology for the errors that you can make are

TABLE 1.1
Confusion Matrix for a Binary
Classifier

Predicted	Actual	
	Yes	No
Yes	tp	fp
No	fn	tn

false discoveries and omissions. False positives and discoveries contribute to the fp count in Table 1.1; similarly, false negatives and omissions contribute to the fn count. What we have just discussed is a generic interpretation of the counts that make up a confusion matrix. For concreteness, it helps to discuss these entries in a particular practical context.

Deciding whether an individual has a certain disease or not is a basic classification problem in medicine. If you regard a diseased individual as belonging to the *positive* class for this problem, then a false positive (negative) results from a given classification if a healthy (diseased) individual is classified as diseased (healthy). Hence, it follows that fp (fn) is the number of healthy (diseased) individuals in the evaluation sample that were classified as diseased (healthy) by the classifier that you use to facilitate the diagnosis. Once you understand this, it should be clear what tp and tn means in this context.

As an example, consider the confusion matrix for the LM classifier (1.6). This array summary is given in Table 1.2 (the issue of how to obtain such a summary will be taken in the next chapter). Adding all numbers in the table gives you the total number of cases in the evaluation dataset. The fraction of this number on the main diagonal gives you a measure of accuracy for the classifier; equivalently, you can consider the classification error rate given by the fraction that makes up the off-diagonal entries. There are, of course, other performance measures that you can derive from the confusion matrix. These measures will be covered in Section 1.3.3 and Chapter 2.

It is important to note that we implicitly assume that a "Yes" corresponds to a *positive* outcome and this outcome identifies the first row and column of Table 1.1. It is not uncommon to encounter other formats for a binary confusion matrix in the machine learning literature. Figure 1.3 shows two examples of alternative formats. The first array summary is from Nwanganga and Chapple [83, p. 322] and the second is from James et al. [62, p. 148]; we leave it to the reader to explore the differences in Exercise 1. Failure to keep in mind the different formats can potentially lead to incorrect identification of the key counts tp, fn, fp, and tn and hence wrong values for performance measures derived from those counts.

Performance measures for binary classifiers are often defined in terms of the key counts. The entries in a confusion matrix that determine these counts depend on the format used for the array summary. The same is true when a column/row profiles approach is used to obtain performance measures (we

TABLE 1.2
Confusion Matrix for the LM Classifier

Predicted	Actual	
	Yes	No
Yes	118	14
No	32	136

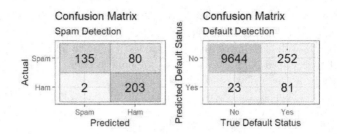

FIGURE 1.3
Confusion Matrices with Formats Different from Table 1.1

discuss this later). Therefore, it is important to be clear about the format used. Our preferred format for binary confusion matrices in this book is given by that shown in Table 1.1.

There is a simpler alternative to the array summary displayed in Table 1.1 that we will often use in this book when discussing performance for binary classifiers. This alternative is the following key count vector:

$$kcv = (tp, fn, fp, tn). \tag{1.9}$$

This 1×4 vector provides a more compact summary of the required information for binary CPA when you need to compute certain performance measures. Note the order of the key counts on the right-hand side of (1.9). We will follow this order whenever we use or make reference to a key count vector.

Thus, instead of the default detection confusion matrix in Figure 1.3, it suffices to provide the key count vector `kcv` shown below.

```
# Key Count Vector for Default Detection Problem

kcv <- c(81, 252, 23, 9644)
names(kcv) <- c("tp", "fn", "fp", "tn")
kcv
##    tp   fn   fp    tn
##    81  252   23 9644
```

To obtain `kcv`, we have adopted the usual practice in default detection of regarding individuals whose default status is "Yes" as belonging to the *positive* class. You need to establish this before you can correctly identify the key counts. Once you have these counts, you can obtain the default detection confusion matrix in the same format as that for Table 1.1. This is easily done with help from the `con_mat()` function that is given in Appendix A.3.

```
# Alternative Default Detection Confusion Matrix

kcv %>% con_mat(c("Yes", "No")) # "conf_mat" object
```

TABLE 1.3

Key Count Vectors for Totally Useless, Random and Perfect Classifiers

Type of Classifier	Key Count Vector (tp, fn, fp, tn)
Totally Useless	$(0, m, n, 0)$
Random	$(m\theta, m(1 - \theta), n\theta, n(1 - \theta))$
Perfect	$(m, 0, 0, n)$

```
##           Actual
## Predicted  Yes    No
##       Yes   81    23
##        No  252  9644
```

We will later need to interpret performance measures that are obtained when a binary classifier is applied to cases in an evaluation dataset. To facilitate the interpretation of such measures, consider three classifiers with key count vectors given in Table 1.3 (performance measures for the corresponding classifiers will be given in the next chapter).

In Table 1.3, m and n are the number of *positive* and *negative* cases in the evaluation dataset, respectively. For a given pair (m, n), the confusion matrix for a totally useless (perfect) classifier is determined by the first (third) key count vector in the table. Clearly, the diagonal (off-diagonal) counts in the corresponding confusion matrix are all equal to 0 for a totally useless (perfect) classifier. Note that having $(m, 0, 0, n)$ as the key count vector is a necessary but not sufficient requirement for a perfect classifier. As we'll see later, this is because there may be other measures (i.e, those not defined by key counts) that do not achieve the corresponding ideal values.

The random classifier is one which assigns a case at *random* to the *positive* class $100\theta\%$ of the time (for a given fractional value θ). Given a particular triple (m, n, θ), the second key count vector in Table 1.3 yields the *expected* confusion matrix for such a classifier that is displayed in Table 1.4.

In practice, when working with a software like **R**, the object used to represent a confusion matrix need not be of class `"matrix"`. Often, it is more convenient to use a `"table"` representation or some other ones like `"conf_mat"`

TABLE 1.4

Expected Confusion Matrix for a Random Classifier

	Actual	
Predicted	Yes	No
Yes	$m\theta$	$n\theta$
No	$m(1 - \theta)$	$n(1 - \theta)$

as is done when using a package like **yardstick** (more on this later). We can demonstrate this by using our `con_mat()` function to construct these objects using the key counts from Table 1.2.

```
# Confusion Matrix as "matrix" and "table" Object

c(118, 32, 14, 136) %>% con_mat(c("Yes", "No"), type = "matrix")
##      Yes  No
## Yes 118  14
## No   32 136

c(118, 32, 14, 136) %>% con_mat(c("Yes", "No"), type = "table")
##          Actual
## Predicted Yes  No
##       Yes 118  14
##       No   32 136
```

The result from the first command is a `"matrix"` object. It seems to be the obvious one to use since it is, afterall, for a confusion matrix. However, we find the `"table"` object from the second command to be more useful for a number of reasons which will become clear in due course (note there is one obvious difference). To obtain a `"conf_mat"` object, leave out the `type` argument in the call to `con_mat()`. Also, note that the `"table"` and `"conf_mat"` representation look the same, but they can produce different results depending on the function you subsequently apply.

The choice of object representation is important because what you get when applying functions like `prop.table()` or `summary()` depends on the object that you supply as argument to these functions. In general, the applicable representation is determined by the task at hand and the **R** package that is being used.

1.3.3   Binary Performance Measures

In practice, classifiers are evaluated using descriptive numerical measures and suitable performance curves. Such evaluations involve the use of a trained classifier and an evaluation dataset to obtain the required numerical and graphical summaries. For the evaluation dataset, use is often made of test data as mentioned in Section 1.3.1, but it can also be one of several resamples used for hyperparameter tuning (we discuss this in Chapter 6). Use of a training dataset for performance evaluation is not recommended because their use tends to produce over-optimistic evaluations.

In terms of the counts in Table 1.1 or the corresponding key count vector (1.9), the measure defined by

$$accuracy = \frac{tp + tn}{tp + fn + fp + tn} \tag{1.10}$$

yields an overall measure of classification accuracy. An equivalent alternative is the overall *error rate* measure given by $1 - accuracy$. However, *accuracy* is more widely used to assess a classifier's ability to produce correct classifications. This is possibly due to the preference for measures that give higher values for better performance.

The other two key measures that we will focus on in this section include the *true positive rate*

$$tpr = \frac{tp}{tp + fn}, \tag{1.11}$$

and the *true negative rate*

$$tnr = \frac{tn}{fp + tn}. \tag{1.12}$$

As defined, *tpr* (*tnr*) is the observed fraction of *positive* (*negative*) cases that are correctly classified. For the medical diagnosis example that we discussed earlier in connection with interpretation of the key counts from a binary confusion matrix, *tpr* (*tnr*) refers to the fraction of diseased (healthy) individuals that were correctly diagnosed by the classifier, given the features recorded for the individuals in the evaluation sample. In the classification literature, depending on the substantive application, it is possible to encounter other terms for class-specific measures of accuracy. For example, *sensitivity* and *specificity* are terms used to refer to *tpr* and *tnr*, respectively, in disciplines like statistics, pattern recognition and epidemiology [90]. In some applications, *tpr* is referred to as *recall*. When discussing CPA measures *in general*, it is probably more meaningful to refer to *tpr* rather than the two alternatives just mentioned because it makes you think of the *true positive rate*. Using application specific terminology can sometimes be counterproductive, especially when used outside the application area.

Like *accuracy*, (1.11) and (1.12) are fractional measures. Hence, values for these measures lie in the interval $[0, 1]$. In the ideal situation when you have zero classification errors (i.e., when $fn = fp = 0$), these measures are all equal to 1. This is hard to achieve in practice when working with realistic classification problems. So, values close to this ideal are generally preferred for these measures. For a perfectly balanced problem (i.e., when $m = n$), $accuracy = 0.5$ for a random classifier but *tpr* and *tnr* depend on θ; see Table 1.5.

TABLE 1.5

Performance Measures for Totally Useless, Random and Perfect Classifiers

Type of Classifier	*accuracy*	*tpr*	*tnr*
Totally Useless	0	0	0
Random	$((m - n)\theta + n)/(m + n)$	θ	$1 - \theta$
Perfect	1	1	1

For the LM classifier given by (1.6), the value for *accuracy* shows that about 85% of cases in the test dataset were correctly classified by the classifier. The *tpr* value shows that about 3 out of 4 *positive* cases in the dataset were correctly classified, and the *tnr* value shows that about 91% of the *negative* cases were correctly classified. Thus, the LM classifier did an excellent job at classifying test data cases in the *negative* class but not quite as well for those in the *positive* class. Overall, its classification accuracy is relatively high.

```
# Key Performance Measures for the LM Classifier

# Key Counts
tp <- 118; fn <- 32; fp <- 14; tn <- 136

# Fraction of Cases that were Correctly Classified
(tp + tn) / (tp + fn + fp + tn) # accuracy
## [1] 0.847

# Fraction in the "Yes" Class that were Correctly Classified
tp / (tp + fn) # tpr
## [1] 0.787

# Fraction in the "No" Class that were Correctly Classified
tn / (fp + tn) # tnr
## [1] 0.907
```

We can infer two false rates from the true rates in the above discussion. These are the *false negative rate* and *false positive rate* defined by the fractional expression in

$$fnr = \frac{fn}{tp + fn} = 1 - tpr \tag{1.13}$$

and

$$fpr = \frac{fp}{fp + tn} = 1 - tnr, \tag{1.14}$$

respectively. The alternative expressions for these false rates follow immediately from the corresponding true rates in (1.11) and (1.12). We can interpret fnr (fpr) as the classification error rate for test data cases in the *positive* (*negative*) class. For the medical diagnosis example discussed earlier, fnr (fpr) is the rate at which a classifier used for the diagnosis incorrectly classifies a diseased (healthy) individual.

Given the true rates that we calculated for the LM classifier, we find that $fnr = 0.213$ and $fpr = 0.093$ for the classifier. Of course, given these numbers, you can also conclude that *sensitivity* $= 0.787$ and *specificity* $= 0.907$ for the classifier. Regardless of your preference for false or true rates, these numbers suggest that the classifier is doing a better job at identifying *negative* cases than *positive ones*.

1.3.4   Binary Performance Curves

Classifier performance measures of scoring classifiers like *tpr* and *fpr* change as we vary the threshold. You can visualize this variation by plotting the receiver operating characteristic (ROC) curve as shown in Figure 1.4 for the LM classifier. In the usual plot of the curve, the vertical axis is determined by *tpr* (i.e., *sensitivity*) and the horizontal axis is determined by *fpr* (i.e., 1 − *specificity*). The curve shows how the fraction of correctly classified *positive* cases varies with the fraction of incorrectly classified *negative* cases as we vary the threshold of the LM classifier.

For a classifier to be useful, its ROC curve should lie above the dashed diagonal line (this line is what you get on average for the ROC curve of a random classifier). This ensures that *tpr* > *fpr* (as it should) no matter what threshold is used for the classifier. The curve should rise up steeply initially and then level off near the horizontal line determined by *tpr* = 1. Ideally, it should look like what you see in Figure 1.5. This ideal is necessary (but not sufficient) for a perfect classifier.

ROC curves have properties that make them especially useful for problems with skewed class distributions and unequal classification error costs [35]. Such curves play an important role when ROC analysis is used in applications like clinical diagnostic testing [82], heart disease detection [44], clinical chemistry [85], bioinformatics [104], radiology [84], clinical psychology [87], and loan pricing [105].

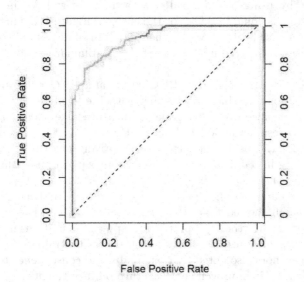

FIGURE 1.4
ROC Curve for the LM Classifier

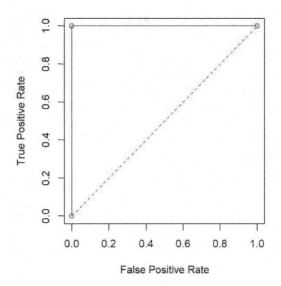

FIGURE 1.5
Ideal ROC Curve

The area under the curve (denoted AUC) is a summary measure that you can obtain from the ROC curve. You can get a value for it by numerical integration, or by using a statistical approach as we will see in the third chapter. This area is equal to 0.934 for the curve in Figure 1.4. This indicates excellent performance for the LM classifier; see Table 4 in Nahm [82] for some guidance on interpreting the magnitude of AUCs. Similar guidance may also be found in [76].

Clearly, $AUC = 1$ for the ideal ROC curve, and $AUC = 0.5$ on average for that of a random classifier. A classifier with the ideal AUC value is not necessarily perfect since other performance measures like *accuracy* may fail to achieve the corresponding ideal value; see Fawcett [35] for an example that illustrates this point. This discrepancy is not surprising since AUC has distinct properties tapping into different aspects of classifier performance [56].

Theoretically and empirically, AUC is a better measure than *accuracy* for evaluating binary classifier performance according to Huang and Ling [61]. Their research showed that AUC is a statistically consistent and more discriminating measure (these properties have precise definitions in their article) than *accuracy*. Optimizing AUC when classifiers are trained also improves *accuracy*. The authors also noted that one real-world consequence to business of such optimization is improvements in return on investments.

Despite the above findings in Huang and Ling [61], the use of AUC as a classifier performance measure has recently been questioned, particularly, when used in a comparative analysis of competing classifiers (its use as a descriptive summary from the ROC curve is not unreasonable for a given classifier whose performance has been established by other coherent measures). Hand [50, 51] highlighted the relatively unknown fact that AUC "is fundamentally incoherent in terms of misclassification costs"; actually, AUC as an inconsistent criterion was noted earlier in a paper by Hilden [60]. Further discussion of the problem with AUC will be taken up in Chapter 3.

1.4 An Overview of Multiclass CPA

Certain binary performance measures like *accuracy* extend naturally to multiclass problems, unlike measures like *sensitivity* or *specificity*. This does not mean that class-specific measures like the last two are not applicable to multiclass problems. For them to be useful, you need to think of the problem as a collection of binary classification problems in which each level of the target variable in turn plays the role of the *positive* class. This view of a multiclass problem is called One-versus-Rest (OvR). With such a perspective, you can compute an OvR collection of binary measures and then get a corresponding measure for the multiclass problem by taking a suitable average.

1.4.1 Required Information

The basic information required for performance evaluation of a multiclass classifier is similar to what is required for a binary classifier, i.e., you need information on predicted and actual classes, and predicted class membership probabilities.

```
# Predictions with Test Data and the ML Classifier

## # A tibble: 300 x 5
##    prob_A      prob_B     prob_C pred_class group
##    <dbl>       <dbl>      <dbl>  <fct>      <fct>
## 1  0.999 0.000681    1.34e- 9 A          A
## 2  1.00  0.00000238 1.19e-12 A          A
## 3  1.00  0.000217    3.62e-10 A          A
## # ... with 297 more rows
```

The above (partial) listing of required information was obtained when the ML classifier given by (1.8) was applied to cases in a test dataset. Here, `pred_class` and `group` have the same interpretations as before. As seen in the earlier binary CPA example, they may be used to obtain descriptive performance measures for the ML classifier. The predicted class membership probabilities for each class is given by `prob_A`, `prob_B` and `prob_C`. For example,

TABLE 1.6

Confusion Matrix for a k-Class Classifier

		Actual		
Predicted	1	2	$\cdots$	k
1	n_{11}	n_{12}	$\cdots$	n_{1k}
2	n_{21}	n_{22}	$\cdots$	n_{2k}
$\vdots$	$\vdots$	$\vdots$	$\vdots$	$\vdots$
k	n_{k1}	n_{k2}	$\cdots$	n_{kk}

`prob_A` contains the predicted probabilities that cases in the test dataset belong to class "A". Together with information in `group`, these probabilities may be used to obtain some performance curves for the classifier.

1.4.2   Multiclass Confusion Matrix

The confusion matrix for a k-class classifier with $k > 2$ is a $k \times k$ array containing counts from a cross-tabulation of predicted versus actual classes in the evaluation dataset. This array summary is given in Table 1.6 when classes are labeled by numbers in $\{1, 2, \ldots, k\}$. In this table, n_{ij} refers to the number of cases in the j-th class of the evaluation dataset that were classified as belonging to the i-th class. Clearly, on setting $k = 2$, we get Table 1.1 after suitable relabeling of classes and cell counts (e.g., relabel n_{11} as tp, etc.). For convenience, we assume labeling of classes is such that the n_{11} count is determined by the reference class.

The confusion matrix for the 3-class ML classifier (1.8) is given in Table 1.7. This was obtained using information from `pred_class` is `group` in the tibble given in Section 1.4.1. Here, labeling of the classes is determined by the levels we assumed for the `group` variable in the simulated dataset; see Figure 1.2 (for convenience, we consider level "A" of the `group` variable as the reference level). In practice, the classes are determined by the substantive problem; hence, more informative labeling may be used for the rows and columns of the confusion matrix. An example is given in Figure 5.5 in Chapter 5.

You can associate a collection of binary confusion matrices with each multiclass confusion matrix by taking an OvR perspective. With such a perspective,

TABLE 1.7

Confusion Matrix for the 3-Class ML Classifier

		Actual	
Predicted	A	B	C
A	74	4	0
B	1	143	8
C	0	3	67

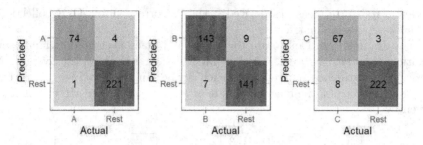

FIGURE 1.6
OvR Collection of Confusion Matrices for the ML Classifier

you can think of a k-class problem in terms of a collection of k binary classification problems. In the i-th binary problem, the i-th class plays the role of the *positive* class, and the remaining classes together constitute the *negative* class. The resulting collection of binary problems play an important role in the definition of some estimates of multiclass performance measures.

The OvR collection of binary confusion matrices is obtained from the multiclass confusion matrix. For example, you can extract the collection shown in Figure 1.6 from the ML confusion matrix that is displayed in Table 1.7. The *positive* class for the first binary confusion matrix in the collection is determined by level "A" of the `group` variable. That for the second and third array summaries are determined by levels "B" and "C", respectively.

1.4.3 Multiclass Performance Measures

In this section, we provide a brief discussion on two key performance measures for multiclass classifiers. Further discussion of multiclass measures will be taken up in Chapter 5.

To begin, as noted earlier, not all performance measures for binary classifiers have a straightforward extension for multiclass classifiers. One exception is the *accuracy* measure. It is still the proportion of cases in the evaluation dataset that determine the counts on the diagonal of a confusion matrix. For the ML classifier, the overall measure of accuracy is given by

$$\frac{74 + 143 + 67}{300} = 0.947.$$

This shows that about 95% of cases in the test dataset were correctly classified. This is one indication of excellent performance.

Measures like *tpr* and *tnr* do not have such straightforward extensions. Without taking an OvR perspective, the notion of *positive* and *negative* class does not make sense for a multiclass problem. However, when you take such a perspective, you can obtain a multiclass *sensitivity* measure by averaging the binary *sensitivity* measures derived from the corresponding OvR collection of binary confusion matrices.

For example, you can obtain the macro and macro-weighted estimates of multiclass *sensitivity* for the 3-class ML classifier by taking the average of *sensitivity* estimates derived from each array summary in Figure 1.6. By using equal weights, the macro estimate of *sensitivity* for the ML classifier is given by

$$\frac{1}{3} \times \frac{74}{74+1} + \frac{1}{3} \times \frac{143}{143+7} + \frac{1}{3} \times \frac{67}{67+8} = 0.944.$$

When weights are derived from prevalence of each class (also known as class priors, i.e., fraction of cases in each class) in the evaluation dataset, you obtain the macro-weighted estimate of the performance measure. Such an estimate of multiclass *sensitivity* for the ML classifier is given by

$$\frac{75}{300} \times \frac{74}{74+1} + \frac{150}{300} \times \frac{143}{143+7} + \frac{75}{300} \times \frac{67}{67+8} = 0.947.$$

See Kuhn and Silge [69, p. 119] for another approach that you can take to obtain the above estimates.

For a perfect 3-class classifier, what values do you expect for the multiclass measures mentioned above? The confusion matrix for such a classifier and all the binary array summaries in the corresponding OvR collection have a diagonal structure. It follows that *accuracy* and both estimates of multiclass *sensitivity* are equal to 1 for such a classifier.

1.4.4   Multiclass Performance Curves

There are several approaches to multiclass ROC analysis. The class reference formulation [35] is one commonly used approach to performance curves for multiclass classifiers. This involves taking an OvR perspective and getting a collection of ROC curves for the multiclass classifier. Such a collection for the 3-class ML classifier is shown in Figure 1.7

Each curve in Figure 1.7 is close to the ideal ROC curve for a trained binary classifier. The *AUC*s for the curves are 1, 0.985, and 0.987, and following Fawcett [35], you can obtain the multiclass *AUC* by macro-weighted averaging

$$\frac{75}{300} \times 1 + \frac{150}{300} \times 0.985 + \frac{75}{300} \times 0.987 = 0.989.$$

The performance measures and curves obtained so far show that the ML classifier is doing an excellent job at solving the multiclass problem.

Other techniques for multiclass ROC analysis like the approach to multiclass *AUC* in Hand and Till [56] and volumes under ROC hypersurfaces will be discussed in Chapter 5.

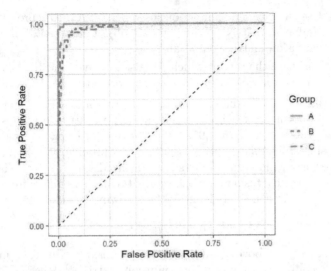

FIGURE 1.7
ROC Curves by Class Reference Formulation

1.5 Exercises

1. As noted earlier in Figure 1.3, confusion matrices can have different formats. In this exercise, we explore potential consequences of using different formats for these array summaries.

   (a) In what way do the formats of the confusion matrices in Figure 1.3 differ from that in Table 1.1?

   (b) Identify the key count vectors for each confusion matrix in Figure 1.3.

   (c) Obtain estimates of *accuracy*, *tpr* and *tnr* and provide a substantive interpretation of the performance measures you computed.

2. The key count vector $(5134, 2033, 6385, 25257)$ is from a confusion matrix given in Kuhn and Johnson [68, p. 39]. The classification problem they considered was concerned with the problem of determining whether a person's profession is in one of the following disciplines: science, technology, engineering, or math (STEM).

   (a) Define the *positive* class for the classification problem. Explain the reasoning for your answer.

(b) Use the con_mat() function in Appendix A.3 to reproduce the confusion matrix. What is the **R** object representation of the confusion matrix you obtain?

(c) Estimate the rate at which (i) false positives and (ii) false negatives occur. What is the overall error rate?

3. Nawanganga and Chapple [83, p. 238] gave an example on application of a k-nearest neighbors classifier for a heart disease classification problem; see Rhys [95, p. 56] for a good discussion on k-NN classifiers. The substantive issue for this problem is to determine whether an individual suffers from heart disease or not. The format for the confusion matrix given by the authors differs from ours. However, we can easily identify the following key counts:

$$tp = 31, fn = 9, fp = 5 \text{ and } tn = 30.$$

(a) Use the above counts and the con_mat() function in Appendix A.3 to obtain the confusion matrix in the same format as used in Table 1.1. Use a "table" object representation.

(b) Obtain the column profiles of the resulting "table" object by dividing entries in each column by the corresponding column total. Interpret the entries on the diagonal.

(c) What is the fraction of correct *positive* classifications (i.e., what fraction of individuals classified as having heart disease actually have the disease)? What is the fraction of correct *negative* classifications?

4. In this exercise, a comparison of two classifiers that were trained using data displayed in Figure 1.1 will be made; both classifiers were trained using age as the only predictor. When the classifiers were applied to test data, the key count vectors shown in the following tibble were obtained for a logit model (LM) classifier and a decision tree (DT) classifier (the next chapter has some information on this machine learner).

```
##    classifier    tp     fn     fp     tn
##    <chr>       <dbl> <dbl> <dbl> <dbl>
## 1 LM            118     32     14    136
## 2 DT            104     46      9    141
```

Note that the key counts for the LM classifier are from Table 1.2.

(a) Which classifier has a smaller overall error rate?

(b) Which classifier has a smaller false positive rate?

(c) Which classifier has a larger true positive rate?

Answer the above questions by (i) inspecting the relevant key counts and (ii) computing relevant performance measures.

5. The following list of confusion matrices is based on what was given in James et al. [62]; their row and column names were changed to Predicted and Actual.

```
## [[1]]
##            Actual
## Predicted  No Yes
##        No  873  50
##       Yes   68   9
##
## [[2]]
##            Actual
## Predicted  No Yes
##        No  920  54
##       Yes   21   5
##
## [[3]]
##            Actual
## Predicted  No Yes
##        No  930  55
##       Yes   11   4
```

In their discussion, the authors considered the classification problem that arises from the question: "Will an individual purchase a caravan insurance policy?". The classifier under consideration in their discussion is based on k-nearest neighbors with $k = 1, 3, 5$; the corresponding confusion matrices are given above (read from top to bottom). The problem under consideration is the question of how performance varies with k. Answer the following for each k.

(a) What is the fraction of *positive* classifications (these are those classified as a "Yes" response) that are correct?

(b) For those who have a *positive* response to the question, what is the fraction of correct classifications?

(c) What is the overall fraction of correct classifications?

Comment on the results that you obtained.

6. For this problem, consider the following 3-class confusion matrix.

```
##            Actual
## Predicted  A  B  C
##        A  20  1  0
##        B   3 10  2
##        C   0  0  8
```

(a) Obtain the OvR collection of binary confusion matrices. Hence, determine the macro and macro-weighted estimates of *specificity*.

(b) Obtain the column profiles of the given multiclass confusion matrix. Average the entries on the diagonal of the resulting matrix and comment on the value you obtained.

2

Classifier Performance Measures

In the last chapter, we provided a glimpse of what is involved when you attempt to evaluate performance of a trained binary classifier. We saw the key role played by the confusion matrix that is obtained from predicted and actual classes of cases in a test dataset. This useful array summary allows you to derive several descriptive performance measures like *accuracy*, *sensitivity*, and *specificity*.[1] These are essentially measures of the number of correct classifications as a fraction of the entire test dataset or similarly defined fractions for each class in this dataset. They are examples of threshold performance measures; see Section 2.3 for further discussion of this important category of measures.

Other relevant measures and issues have been examined and reported in the literature. For example, the review by Tharwat [108] considered several other performance measures and examined the influence of balanced and imbalanced data on each metric. As part of their study, Sokolova and Lapalme [103] considered the issue of reliable evaluation of classifiers by examining the invariance properties of several performance measures. Concepts and measures such as informedness and markedness that reflect the likelihood that a classification is informed versus chance was examined by Powers [88]. Recently, Aydemir [3] proposed the polygon area metric (PAM) to facilitate CPA when there are competing classifiers for a given problem. Given the wide variety of performance measures, it should come as no surprise that different measures quantify different aspects of performance (this is a noteworthy point); see the excellent review by Hand [52] for further discussion on this issue and other aspects like choice of and comparisons between performance measures.

In this chapter, we extend coverage of threshold performance measures to include, among others, those that help to answer the third and fourth questions in Section 1.3. Another important aim is to show how **R** may be used to analyze classifier performance. In particular, we demonstrate functionality in the **yardstick** package for this purpose. Some of the measures we cover may also be viewed as estimates of certain performance parameters. For these metrics, we also discuss some procedures that can be used to make inferences

---

[1]Equivalent measures that are complementary to the listed ones include *error rate*, *false negative rate*, and *false positive rate*.

on them. To illustrate discussion, we use a decision tree classifier that is trained using data from the **titanic** package.

2.1 Classification with Machine Learners

You are not restricted to the use of statistical learners to solve classification problems. With increased computing power, you can also take an algorithmic approach to construct machine learning classifiers like the decision tree in Figure 2.1 or the neural network in Figure 3.1. The recursive partitioning algorithm is one approach you can use to train the former, and the backpropagation algorithm is one option you have for training the latter.

2.1.1 Decision Tree Classifier

Decision tree (DT) classifiers have great appeal even though they may not necessarily be the best machine learner for a given classification problem. Some reasons for this include the fact that little preprocessing is required during training (e.g., categorical features do not need preprocessing), outliers have little impact on results, missing values do not require imputation, and the classification rules associated with the classifier may be displayed as an inverted tree to aid understanding. Further discussion on strengths and weaknesses of tree-based algorithms may be found in Nwanganga and Chapple [83, p. 298], for example.

To see what is involved in a tree-based approach to classification, consider the decision tree in Figure 2.1 that was constructed using data that is displayed

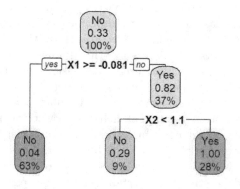

FIGURE 2.1
Decision Tree from Simulated Training Data

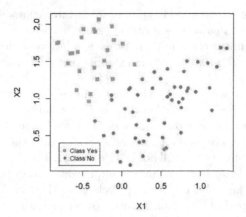

FIGURE 2.2
Simulated Training Data

in Figure 2.2 (in this dataset, `group` is the target variable with two levels and the features are the numeric variables X1 and X2).

The inverted tree structure in Figure 2.1 is typical of a decision tree; it is made up of three types of nodes that are linked by branches. The root node is at the top and the terminal nodes are at the bottom. The nodes in between these two types are called internal nodes. For the given decision tree, we have one root node, one internal node, and three terminal nodes.

The root node of a decision tree contains all the data prior to splitting. A suitable criterion is used to split this node into two branches, each of which leads to another node. Internal nodes are also split in a similar fashion. Nodes that do not split further make up the terminal nodes. The whole process is governed by a recursive partitioning algorithm like **rpart**. This version is an open source implementation of **CART**, the classification and regression trees algorithm developed by Breiman et al. [9].[2]

A measure of impurity like *Gini index* that quantifies how heterogeneous classes are within a node may be used in the node-splitting process. This measure is defined by

$$Gini\ index = \sum_{i \in C} p_i(1 - p_i) = 1 - \sum_{i \in C} p_i^2,$$

where p_i is the proportion of observations belonging to class i, and it summarizes total variance across the classes. It takes on a small value if all of the p_i's are close to zero or one. Thus, a small value of this index indicates that a node contains predominantly observations from a single class (indicating node

---

[2]The **C4.5** algorithm by Quinlan [92] is also popular; see Roiger and Geatz [97, p. 68] for a tutorial on this algorithm.

purity is high). The feature used to split a root or internal node is the one with the best value for *Gini gain* (this is the difference between the *Gini index* of a parent node and a linear combination of that for the two nodes resulting from the split). Rhys [95, p. 170] has a good example that demonstrates calculation of this criterion.

To classify a case with the DT classifier, start at the root node and move down along each branch (take the left branch if the condition associated with a node is true) until you reach a terminal node. The label in the latter determines the class of the case. For example, a case is assigned to the "Yes" class if it ends up in the right-most terminal node in Figure 2.1. The recorded percentage shows that 28% of cases in the training sample winds up in this node. A case that winds up in this node is certain to belong to the "Yes" class as shown by the estimated probability (i.e., number below the label for the node) of belonging to the "Yes" class given the node conditions satisfied by the case.

Finally, note that decision trees tend not to perform well due to the tendency to overfit. Hence, they tend to have high variance (a small change in the training data can produce big changes in the fitted tree and hence highly variable predictions). To guard against extravagant tree building, you can employ suitable stopping criteria (e.g., minimum number of cases in a node before splitting, maximum depth of the tree, minimum improvement in performance for a split, minimum number of cases in a leaf). Alternatively, you can apply suitable pruning techniques to reduce the size of the decision tree; see Boehmke and Greenwell [6, p. 181], for example.

2.2 Titanic Survival Classification Problem

For convenience, we will use the problem in the heading for this section as a running example in this and the next two chapters. Rhys [95, p. 98] provided one possible motivation for interest in this classification problem. Also given in this reference is a description of the variables in the dataset that will used and some exploratory analysis of the available data. The steps used to prepare data in the next section follow in part with what was done in Rhys [95, p. 101].

2.2.1 Data Preparation

The dataset we will use is the `titanic_train` data frame from the **titanic** package. After some preprocessing, we will split this dataset into two parts; one part to be used for training the DT classifier and the other to be used for evaluating its performance.

Note the use of **tidyverse** in the **R** code for data preparation. Several core packages (e.g., **dplyr**, **ggplot2**, and **purrr**; see Appendix A.1 for more information) in this excellent meta-package are used in this and subsequent chapters. You should load it in your **R** session if you want to run the code in this book.

```
# Data Preparation

library(tidyverse) # load several core packages

# Get titanic_train from the titanic Package
data(titanic_train, package = "titanic")

# Create the Titanic_df Data Frame
Titanic_df <-
    titanic_train %>% # see R20-99
    mutate(FamSize = SibSp + Parch) %>%
    select(Survived, Sex, Age, FamSize, Pclass) %>%
    mutate_at(.vars = c("Survived", "Sex", "Pclass"),
        .funs = factor) %>%
    mutate(Survived = fct_recode(Survived, "Yes" = "1",
        "No" = "0")) %>%
    mutate(Pclass = fct_recode(Pclass, "1st" = "1", "2nd" = "2",
        "3rd" = "3"))

# Check for Missing Values
Titanic_df  %>% map_int(~ is.na(.) %>% sum())
## Survived       Sex      Age  FamSize    Pclass
##        0         0      177        0         0

# Impute Missing Values
Titanic_tb <- Titanic_df %>%
    replace_na(list(Age = mean(Titanic_df$Age, na.rm = TRUE))) %>%
    as_tibble()
Titanic_tb %>% print(n = 3)
## # A tibble: 891 x 5
##    Survived Sex       Age FamSize Pclass
##    <fct>    <fct>   <dbl>   <int> <fct>
## 1 No        male       22       1 3rd
## 2 Yes       female     38       1 1st
## 3 Yes       female     26       0 3rd
## # ... with 888 more rows
```

The `Titanic_df` data frame created in the above code segment has some missing values.[3] For convenience, we impute the missing values using the

---

[3]Note the use of the basic pipe operator `%>%` from the **magrittr** package. We make frequent use of this and other pipe operators (e.g., `%T>%` and `%$%`) in this book.

`replace_na()` function from the **tidyr** package so that there is a complete dataset to start with. There is, of course, a better way to deal with the missing data in practice. You can, for instance, incorporate the required imputation in a preprocessing pipeline as part of the workflow when you use the **tidymodels** meta-package to train and test your classifier as shown in Section 6.2.2. The difference does not matter here because our aim is illustrate various techniques for classifier performance analysis. For this purpose, it suffices to have a trained classifier and a test dataset.

2.2.2   Training a DT Classifier

As noted in the last chapter, training a classifier is quite an involved process that you need to go through if you want a trained classifier that performs well for your practical problem. However, for the DT classifier that we are about to train, we'll keep things simple and concentrate our efforts on training data preparation and model fitting (without hyperparameter tuning). We do this because our focus in this chapter is to illustrate the use of performance measures to evaluate a binary classifier and not on getting the best classifier for the problem at hand.

We use 75% of the cases in the `Titanic_tb` tibble from the last section to make up the required training dataset. You can change the `prop` argument to the `initial_split()` function (from the **rsample** package) to specify a different proportion of data for training (the default proportion is 0.75). The remaining 25% of cases in `Titanic_tb` constitute the test dataset. It is important to note that this dataset should never be used when training a classifier.

```
# Training Dataset for the DT Classifier

# Create the Training Dataset
library(rsample) # core package in tidymodels
set.seed(19322)
Titanic_split <- Titanic_tb %>% initial_split()
Titanic_train <- Titanic_split %>% training()

# Partial Listing of the Training Dataset
Titanic_train %>% print(n = 3)
## # A tibble: 668 x 5
##    Survived Sex      Age FamSize Pclass
##    <fct>    <fct> <dbl>   <int> <fct>
## 1 No       male  20         0 3rd
## 2 No       male  20         0 3rd
## 3 No       male  29.7      10 3rd
## # ... with 665 more rows
```

Next, we use the `rpart()` function in the **rpart** package to fit the decision tree. There are hyperparameters associated with this function that govern the

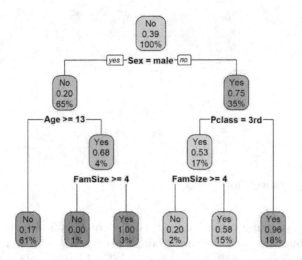

FIGURE 2.3
DT Classifier for the Titanic Survival Classification Problem

tree construction process. These include those that determine tree complexity, maximum depth and minimum number of cases in a node for it to be split further. For the moment, we use default values for these hyperparameters.

```
# DT Classifier for Titanic Survival Classification

# Fit the Decision Tree
library(rpart)
DT_fit <- rpart(Survived ~ ., data = Titanic_train)

# Display the DT Classifier
library(rpart.plot)
DT_fit %>% rpart.plot(box.palette="BuRd")
```

Figure 2.3 provides a display of the fitted decision tree. You can also refer to it as a display of the trained DT classifier for the Titanic survival classification problem since the figure is a display of the rule set associated with the DT classifier for this problem.[4]

As can be seen from the displayed decision tree, an individual starting from the root node will reach the left-most leaf node at the bottom of the decision tree if $Sex = male$ and $Age \geq 13$ are both true for the individual. The numbers in this leaf node show that 61% of individuals in the training

---

[4]See Roiger and Geatz [97, p. 11] for a simple example that shows how a decision tree can be translated to a set of production rules.

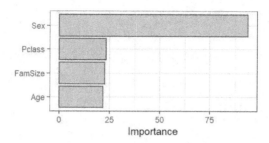

FIGURE 2.4
Variable Importance Plot

dataset satisfied these conditions and they were classified as non-survivors. The corresponding conditional probability of survival is 0.17. Individuals that wind up in right-most leaf node constitute 18% of individuals in the training dataset, and these individuals were classified as survivors. It also shows that there is an estimated 0.96 chance of survival for females in the first or second class.

The gender of an individual is the most important feature when it comes to predicting the survival status of an individual. This is demonstrated in two ways, i.e., we see that *Sex* is the feature selected to split the root node and its importance far exceeds the other features as can be seen in the variable importance plot; see Figure 2.4.

```
# Variable Importance Plot (VIP)

# Display the VIP
library(vip)
DT_fit %>% vip(aes = list(color = "blue", fill = "lightblue"))
```

2.2.3   Predictions from the DT Classifier

To assess performance of a trained classifier, you need to use it and data in an evaluation (e.g., test) dataset to obtain information on predicted probabilities and classes. Recall from our discussion in the last chapter, that predicted class membership probabilities are required to construct some of the commonly used performance curves and predicted classes are required for calculation of descriptive performance measures. In both situations, you also need data on actual classes from the evaluation dataset. For our purpose in this chapter, we'll use the `Titanic_test` dataset created below to evaluate the trained DT classifier displayed in Figure 2.3.

```
# Test Dataset for the DT Classifier

# Create the Test Dataset
Titanic_test <- Titanic_split %>% testing()

# Partial Listing of the Test Dataset
Titanic_test %>% print(n = 3)
## # A tibble: 223 x 5
##    Survived Sex       Age FamSize Pclass
##    <fct>    <fct>   <dbl>   <int> <fct>
## 1 Yes      female     26       0 3rd
## 2 No       male       35       0 3rd
## 3 No       male        2       4 3rd
## # ... with 220 more rows
```

The required predictions may be obtained as shown below. To evaluate performance of the trained DT classifier, it suffices to have information in the last three columns of the DT_pv tibble. Both prob_Yes and pred_class have the same interpretations as in Section 1.3.1. Here, Survived contains information on the actual classes.

```
# Test Data Predictions from the DT Classifier

# Obtain Predicted Probabilities & Classes
DT_pv <-
  bind_cols(
      predict(DT_fit, newdata = Titanic_test, type = "prob") %>%
         as_tibble() %>% setNames(c("prob_No", "prob_Yes")),
     tibble(
        pred_class =  predict(DT_fit, newdata = Titanic_test,
           type = "class"),
        Survived = Titanic_test %$% Survived
     )
  )

# Partial Listing of the Predictions
DT_pv %>% print(n = 3)
## # A tibble: 223 x 4
##    prob_No prob_Yes pred_class Survived
##      <dbl>    <dbl> <fct>      <fct>
## 1   0.423    0.577 Yes        Yes
## 2   0.835    0.165 No         No
## 3   1        0     No         No
## # ... with 220 more rows
```

When working with functions in the **yardstick** package for CPA, keep in mind that by default the first level of the target variable is the reference level. If you want the second level to be the reference in binary CPA, you need to specify event_level = "second" as one of the arguments in your call to the performance function when applicable. Henceforth, for convenience, we will make the *positive* class the first level of factor variables that define the predicted and actual classes (since this avoids the need for the extra argument). As shown in the next example, the "Yes" level (which defines the *positive* class for our problem) is the second level of both pred_class and Survived. We can make it the first level by using the factor reversal function fct_rev() in the **forcats** package.

```
# Checking and Changing Reference Levels

# Check Reference Levels
DT_pv %>% keep(is.factor) %>% map(~ levels(.))
## $pred_class
## [1] "No"  "Yes"
##
## $Survived
## [1] "No"  "Yes"

# Make Yes the Reference (i.e., Positive) Class
DT_pv <- DT_pv %>% mutate_if(is.factor, fct_rev)
```

2.2.4   Array Summaries

A cross-tabulation of the predicted versus actual class variables is usually the first summary to obtain after getting results like that in the DT_pv tibble.

```
# Confusion Matrix for the DT Classifier

# By Using the table() Function
DT_pv %$% table(pred_class, Survived,
  dnn = c("Predicted", "Actual"))
##          Actual
## Predicted Yes  No
##       Yes  57  12
##       No   22 132

# By Using the conf_mat() Function in yardstick
library(yardstick) # core package in tidymodels
DT_pv_cm <- DT_pv %>% conf_mat(Survived, pred_class,
    dnn = c("Predicted", "Actual")) # "conf_mat" object
DT_pv_cm %>% autoplot(type = "heatmap")
```

You can obtain the required array summary in a number of ways. As shown in the preceding code segment, the first approach uses the `table()` function and hence yields a `"table"` object (note the use of the exposition pipe `%$%` here). There are situations when it is advantageous to take this approach. The second approach that uses the `conf_mat()` function from the **yardstick** package is also very useful. It yields a `"conf_mat"` object and Figure 2.5 shows a heat map display of this object. Such a display is one of several benefits that you have when `conf_mat()` is used to obtain a confusion matrix.

There is another approach that is not widely known that yields a `"matrix"` object and it involves use of the indicator matrices that we can get for `pred_class` and `group` with information in `DT_pv` (we demonstrate this later when we discuss multiclass CPA). This shows the importance that these indicator matrices play in the determination of CPA measures. In particular, we see in the next section that a measure called Matthews correlation coefficient relies on them. Reliance on these matrices is necessary because the formula that we will see later in this chapter for this measure in terms of key counts is not very useful when it comes to generalization for use with multiclass classifiers. The resulting approach is conceptually and computationally appealing.

The essential information to extract from a confusion matrix are the key counts since they play an important role when you need to compute performance measures that depend on these counts, and you use the formulas that define them. This approach is sometimes convenient to use even though you have a package like **yardstick** to facilitate calculations of the required measures. When attempting to identify the key counts for a given binary problem, note that identification of what constitutes the *positive* class is an important first step. This is easily done with the confusion matrix in Figure 2.5 in light of the format we adopted in Table 1.1 of Chapter 1 for this array summary. You have to be careful when your confusion matrix is not in the adopted format.

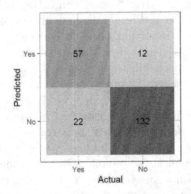

FIGURE 2.5
Confusion Matrix for the DT Classifier

The key counts from Figure 2.5 are contained in the `kcv` vector below.

```
# Key Count Vector for the DT Classifier

tp <- 57; fn <- 22; fp <- 12; tn <- 132 # key counts
kcv <- c(tp, fn, fp, tn) # vector of key counts
names(kcv) <- c("tp", "fn", "fp", "tn")
kcv
##   tp  fn  fp  tn
##   57  22  12 132
```

The key counts in the above code segment allow you to obtain a preliminary assessment of the DT classifier. If required, you can compute an overall measure of classification accuracy and the rates at which false negatives and false positives occur by using these counts and the formulas given by (1.10), (1.13), and (1.14). As we'll see in the subsequent sections of this chapter, there are other useful performance measures that you can also compute.

2.3   Threshold Performance Measures

Descriptive summaries that are used to assess performance of binary classifiers may be obtained in a number of ways. Most of the commonly used measures (e.g., *accuracy*, *tpr*, and *tnr* from the last chapter) are functions of the key counts *tp*, *fn*, *fp*, and *tn* from a binary confusion matrix. The threshold performance measures that we will focus on in this section are determined by these counts. According to the taxonomy proposed by Ferri et al. [37], this group of measures are those that "depend on a threshold and a qualitative understanding of classification errors"; they are applicable when classification error minimization is an important issue to consider. Two other proposed groups include ranking measures (i.e., those that determine effectiveness in separating classes), and probabilistic measures (i.e., those that measure uncertainty in the classifications). Discussion of the last two groups of measures will be taken up later.

Not all performance measures suggested in the literature are useful (e.g., *AUC* is not very useful for comparative performance analysis of competing classifiers), and there are also some redundancies (e.g., *tpr* and *fnr* are linearly related, hence one of them is redundant). Clearly, some criteria are needed to help you select the more useful ones. It is not unreasonable to require a performance measure to be scalar valued, computationally tractable, and have intuitive appeal. Good reasons for this expectation may be found in Anagnostopoulos et al. [2], for example. These authors also mentioned another quality, namely, the need for a measure to coherently capture the aspect of performance of interest; see Hand [50] for an example where this requirement is violated.[5] Other criteria may also be relevant, e.g., measure invariance [103], applicability to imbalanced classification problems [10], and sensitivity to costs of different misclassification errors [36].

Measures like *accuracy* and *tpr* satisfy most of the abovementioned requirements; some issues with these measures will be discussed later. Recall, they provide answers to the first two questions given in Section 1.3. We need to expand our coverage of performance measures to answer the remaining two questions and others that are concerned with classifier performance. The threshold measures we cover may be further divided into three groups, namely, those that are class-specific, those that measure overall performance, and composite measures. Typically, definition of measures in the last two groups involves all of the four key counts and those in the first group usually rely on a subset of these counts.

We take a slightly different approach when we review the measures mentioned above and introduce new ones in this section. The approach assumes familiarity with what is involved in getting the column/row profiles of a rectangular array like Table 1.1. This approach has some advantages like the ability to easily obtain estimates of multiple CPA measures and to obtain them even without relying on the formulas that define the measures, or use functions in a package like **yardstick**. The approach also allows for several relationships among CPA measures to be easily deduced (e.g., the relationship between true and false rates).

2.3.1   Class-Specific Measures

Performance measures like the true rates *tpr* and *tnr* (and related false rates *fnr* and *fpr*) are examples of class-specific threshold performance measures for binary classifiers. These measures focus on how a classifier performs for a particular class. We discussed them in the last chapter and, for the discussion to follow, it is worthwhile restating their definitions:

$$
\begin{aligned}
tpr &= \textit{true positive rate} = tp/(tp+fn),\\
fnr &= \textit{false negative rate} = fn/(tp+fn),\\
fpr &= \textit{false positive rate} = fp/(fp+tn),\\
tnr &= \textit{true negative rate} = tn/(fp+tn).
\end{aligned}
$$

As shown in Table 2.1, the fractions involved in the above definitions appear in the column profiles of the generic confusion matrix that was given in Table 1.1. Thus, the four measures listed above form a natural group of fractional performance measures. A useful practical implication is the fact that you

---

[5]This example is important and we'll return to it later since the discussion involves the area under the ROC curve (a topic that we'll cover with more detail in the next chapter).

TABLE 2.1
Column Profiles of the Confusion Matrix in
Table 1.1

	Actual	
Predicted	Yes	No
Yes	$tp/(tp + fn)$	$fp/(fp + tn)$
No	$fn/(tp + fn)$	$tn/(fp + tn)$

can obtain these measures easily with software like **R** since its `prop.table()` function may be used to calculate column (and row) profiles.[6] As will be seen later, this is not the only advantage when you adopt this perspective.

The *tpr* measure provides an answer to the second question given at the beginning of Section 1.3 (replacing *positive* in the question by *negative* leads to another question whose answer is provided by *tnr*). Thus, *tpr* may be viewed as a measure of classification accuracy for the *positive* class. You can also answer the question just mentioned if you know *fnr* since

$$tpr + fnr = 1.$$

This relationship follows from Table 2.1 since the first (and second) column total for this table is clearly equal to 1. Thus, another advantage of the column profiles perspective is that it allows you to easily deduce relationships between true and false rates.

In practice, some measures are often referred to by different names that make sense in a particular application area. As noted in the last chapter, terms like *sensitivity* and *specificity* are used in certain disciplines instead of *tpr* and *tnr*. A measure like *sensitivity* is important in binary classification problems that arise in medical applications like disease detection. In such applications, a diseased individual belongs to the *positive* class. For some diseases, it is necessary to use a classifier with high *sensitivity* to ensure that false negatives occur rarely or not at all (this is particularly important for harmful diseases with high infection rate).

There are other class-specific measures that you can obtain from a binary confusion matrix. For example, you can consider the measures from the row profiles of Table 1.1. The entries in Table 2.2 provide another set of four performance measures:

$$
\begin{aligned}
ppv &= \textit{positive predicted value} = tp/(tp + fp), & (2.1) \\
fdr &= \textit{false discovery rate} = fp/(tp + fp), \\
for &= \textit{false omission rate} = fn/(fn + tn), \\
npv &= \textit{negative predicted value} = tn/(fn + tn). & (2.2)
\end{aligned}
$$

---

[6]Column (row) profiles of Table 1.1 are obtained by dividing entries in each column (row) by the corresponding column (row) total. Ferri et al. [38] refer to Table 2.1 as the confusion ratio matrix.

TABLE 2.2

Row Profiles of the Confusion Matrix in
Table 1.1

Predicted	Actual	
	Yes	No
Yes	$tp/(tp+fp)$	$fp/(tp+fp)$
No	$fn/(fn+tn)$	$tn/(fn+tn)$

The answer to the third question at the beginning of Section 1.3 is provided by the *ppv* measure. Replacing *positive* in the question by *negative* yields another related question whose answer is provided by *npv*. Also, note that in general, $fpr \neq fdr$ even though the number of false positives and number of false discoveries are determined by fp in Table 1.1. Similarly, $fnr \neq for$ in general. It suffices to examine the denominator in the definition of the relevant rates to see why this is so.

In text classification and information retrieval, *ppv* is also referred to as *precision* [90]. In such applications, *recall* is an alternative term for *sensitivity*. Also, in database record linkage applications, *precision* and *recall* (including their harmonic mean) are popular measures because the underlying classification problem is often very unbalanced [46].

In applications such as spam classification, *precision* takes precedence over *recall*. This is because it is important to have a high fraction of correct classifications when emails are classified as spam, i.e., we want a very high *precision* value. Another way of saying this is to say that we require a very low *false discovery rate* (this measure is equal to $1 - ppv$). However, it is possible for your business goal to be stated in terms of *sensitivity* and *specificity* for a spam classification problem; see the conclusion resulting from the illustrative dialogue process in Table 6.3 of Zumel and Mount [119], for example.

As noted earlier, calculation of binary class-specific measures by the column/row profiles approach may be performed by using the `prop.table()` function. The array summary you supply to this function as one of its arguments must be of class `"table"` or `"matrix"`.

The confusion matrix for the DT classifier was obtained earlier in the last section; see Figure 2.5. This figure is a display of `DT_pv_cm` that we created earlier. It is an object of class `"conf_mat"`, not one of the two that is required. Fortunately, you can use the `pluck()` function from the **purrr** package to extract the required `"table"` object from `DT_pv_cm` before calculating the required profiles. This is demonstrated next.[7]

---

[7] If necessary, refer to Table 2.1 and Table 2.2 for help to interpret the entries in the resulting column and row profiles.

```
# Class Specific Measures from Column/Row Profiles

# Class Specific Measures: tpr, fnr, fpr and tnr
DT_pv_cm %>% pluck(1) %>% prop.table(2) %>% round(3)
##             Actual
## Predicted   Yes    No
##       Yes 0.722 0.083
##        No 0.278 0.917

# Class Specific Measures: ppv, fdr, for and npv
DT_pv_cm %>% pluck(1) %>% prop.table(1) %>% round(3)
##             Actual
## Predicted   Yes    No
##       Yes 0.826 0.174
##        No 0.143 0.857
```

Thus, for the trained DT classifier, the *tpr* (*tnr*) value tells you that the fraction of survivors (non-survivors) in the test dataset that were given the correct classification is 0.722 (0.917). The classifier did a very good job at classifying non-survivors but not as well for survivors. Furthermore, the value obtained for *ppv* (*npv*) from the row profiles show that the fraction of correct survivor (non-survivor) classifications is 0.826 (0.857). Classifying a passenger as a non-survivor is more likely to be correct than classifying a passenger as a survivor.

Figure 2.6 provides a visualization of the column profiles from the confusion matrix for the DT classifier. The relative magnitudes of the relevant rates are easily seen in this figure. For example, you can see that false positives occur at a lower rate than false negatives. Use the following code to produce the figure.

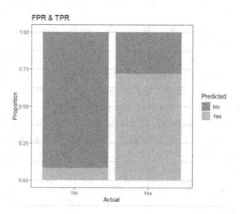

FIGURE 2.6
Bar Chart of False & True Positive Rates

```
# Bar Chart of False & True Positive Rates

DT_pv %>%
  rename(Actual = Survived, Predicted = pred_class) %>%
  mutate_at(.vars = c("Actual","Predicted"), .funs = fct_rev) %>%
  ggplot(aes(x = Actual, fill = Predicted)) + theme_bw() +
  scale_fill_manual(values = c("#CC0000","#5DADE2")) +
  geom_bar(position = "fill") + ylab("Proportion") +
  ggtitle("FPR & TPR")
```

You can use the **vtree()** function to obtain a more informative display of the true and false rates (in percentage terms) from the column profiles; see the variables tree plot in Figure 2.7. The plot is quite informative since the key counts are also displayed in addition to estimates of the class priors (i.e., prevalence of the classes) and the size of the test dataset.[8]

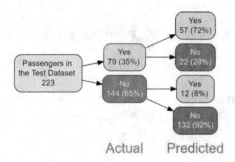

FIGURE 2.7
Variable Tree Display of True and False Rates

```
# Variable Tree Display of True and False Rates

library(vtree)
DT_pv_cm %>% pluck(1) %>%
  crosstabToCases() %>%
  vtree("Actual Predicted", imagewidth="3in", imageheight="2in",
    title = "Passengers in \n the Test Dataset")
```

In practice, functions in a suitable package are used to compute the performance measures. When using the **yardstick** package, you can obtain values for *tpr*, *tnr*, *ppv*, and *npv* from data on predicted and actual classes in DT_pv as shown in the next code segment.

---

[8]A similar plot may also be obtained for class-specific measures from the row profiles.

```
# Class-Specific Measures Using yardstick Functions

library(magrittr) # for the exposition pipe operator

DT_pv %$% sens_vec(Survived, pred_class)
## [1] 0.722
DT_pv %$% spec_vec(Survived, pred_class)
## [1] 0.917
DT_pv %$% ppv_vec(Survived, pred_class)
## [1] 0.826
DT_pv %$% npv_vec(Survived, pred_class)
## [1] 0.857
```

Note that the **magrittr** package was loaded in the above code segment to enable use of the exposition pipe operator %$%.[9] Also, note that for each measure we calculated, there are two functions in the **yardstick** package for it. For example, you can also obtain *sensitivity* as shown below.

```
# Using Alternative yardstick Function for sensitivity

DT_pv %>% sens(Survived, pred_class)
## # A tibble: 1 x 3
##   .metric .estimator .estimate
##   <chr>   <chr>          <dbl>
## 1 sens    binary         0.722
```

Each alternative function returns a tibble instead of a single numerical value. They are useful for some situations when **R** is used to solve problems in CPA. See Appendix A.2 for other alternative functions.

```
# Key Class-Specific Measures Using yardstick

DT_pv_cm %>% summary() %>% slice(3:6)
## # A tibble: 4 x 3
##   .metric .estimator .estimate
##   <chr>   <chr>          <dbl>
## 1 sens    binary         0.722
## 2 spec    binary         0.917
## 3 ppv     binary         0.826
## 4 npv     binary         0.857
```

The above code segment shows you how to easily get a number of performance measures in the same format as that shown in the above alternative

---

[9]For the basic pipe operator %>%, you do not need to do this loading to use it if you have already loaded the **tidyverse** meta-package.

calculation of *sensitivity*. You get the results by supplying your confusion matrix as a "conf_mat" object to the summary() function. The slice() function from the **dplyr** package allows you to select the required measures. See Appendix A.3 for another example that demonstrates the utility of this approach.

2.3.2 Overall Measures

The usual overall threshold measure of performance for binary classifiers is the *accuracy* measure defined by (1.10). Although widely used, the utility of this measure is diminished when you have serious class imbalance (i.e., when the number of cases are significantly different for the two classes), and unequal classification error costs is of concern. Absent these concerns, it is a reasonable measure because of its intuitive appeal. Furthermore, for it to be useful, *accuracy* must be (significantly) greater than the "No Information Rate" (this is determined by the prevalence of the majority class, i.e, the fraction of cases in this class).

```
# accuracy of the DT Classifier

# accuracy
DT_pv %$% accuracy_vec(Survived, pred_class)
## [1] 0.848

# Estimate of the No Information Rate
DT_pv %$% table(Survived) %>% prop.table() %>% max()
## [1] 0.646
```

Imbalanced datasets can pose a problem especially when the minority class is very small. When the dataset is highly imbalanced, the trivial classifier that assigns every case to the dominant class will have a high value for *accuracy*. For example, an *accuracy* of 99.99% can easily be obtained by classifying all record pairs as non-matches when record linkage is performed with large databases [46]. As another example, the trivial classifier for the spam classification problem that assigns every email to the non-spam class will have high *accuracy* when prevalence of spam is very low, i.e., when you have an insignificant fraction of spam emails.

The true rates given by equations (1.11) and (1.12) have the following simple relationship with *accuracy*:

$$accuracy = tpr \times prevalence + tnr \times (1 - prevalence), \qquad (2.3)$$

where *prevalence* is given by $(tp + fn)/(tp + fn + fp + tn)$, i.e., it is the fraction of actual positives in the evaluation dataset. You can easily verify (2.3) by substituting the formulas for the true rates on the right-hand side of this equation.

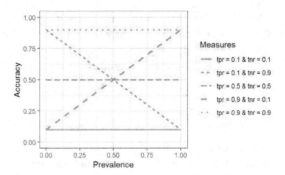

FIGURE 2.8
Accuracy vs Prevalence for Given True Rates

Clearly, *accuracy* varies linearly with *prevalence*. The intercept and slope are given by *tnr* and (*tpr* − *tnr*), respectively. Figure 2.8 displays the linear relationship for five different sets of values for the true rates. The plot shows that *accuracy* is equal to the common value of the true rates when they are equal regardless of *prevalence*. When *tpr* is greater (smaller) than *tnr*, *accuracy* increases (decreases) with *prevalence*. One of the latter alternatives is usually the case.

Note that there is also a simple relationship between *accuracy*, *ppv*, and *npv*. This is given by

$$accuracy = ppv \times dp + npv \times (1 - dp),$$

where *dp* is *detection prevalence* defined by $(tp + fp)/(tp + fn + fp + tn)$, i.e., it is the fraction of cases that are classified as *positive* in the evaluation dataset.

Correct classification by chance alone can also be a problem when you have highly imbalanced data. Cohen's *kappa* provides an adjustment to predictive accuracy by accounting for this potential problem [83]. It does this by measuring the concordance between actual and predicted responses for a classification problem. This measure is given by

$$kappa = \frac{\hat{A} - \hat{E}}{1 - \hat{E}}, \tag{2.4}$$

where, $\hat{A}$ is the *accuracy* measure (1.10) and $\hat{E}$ is an estimate of expected accuracy under chance agreement that is given by

$$\hat{E} = \frac{(tp + fp)(tp + fn) + (fn + tn)(fp + tn)}{(tp + fn + fp + tn)^2}.$$

This formula is a special case of that given in Chapter 5.

Values of the *kappa* measure above 0.5 indicate moderate to very good performance, while values below 0.5 indicate very poor to fair performance

[83]. The measure is a commonly used alternative to *accuracy* when assessing overall performance of a classifier. However, note that it can also produce unreliable results due to sensitivity to the distribution of marginal totals [22]. Fortunately, we have another overall CPA measure called Matthews correlation coefficient (*mcc*) that is given by

$$mcc = \frac{tp \times tn - fp \times fn}{\sqrt{(tp+fn)(fp+tn)(tp+fp)(fn+tn)}}. \qquad (2.5)$$

This measure has its origins in bioinformatics, an application area where class imbalance occurs very often. Derivation of (2.5) may be found in Delgado and Tibau [22, p. 5], for example.

When Matthews [78] considered *mcc* in his study, he computed it by finding Pearson's correlation between two indicator vectors. We demonstrate this in the following code segment. Here, the indicator vectors are for the "Yes" level of `pred_class` and `Survived` in the `DT_pv` tibble.

```
# Pearson's Correlation Calculation of mcc

library(ncpen) # for to.indicator()
DT_pv %>%
  select(pred_class, Survived) %>%
  map(~ to.indicators(., exclude.base = FALSE) %>% .[, 1]) %>%
  as_tibble() %T>% print(n = 5) %$%
  cor(pred_class, Survived)
## # A tibble: 223 x 2
##    pred_class Survived
##         <dbl>    <dbl>
## 1           1        1
## 2           0        0
## 3           0        0
## 4           1        1
## 5           0        0
## # ... with 218 more rows
## # i Use 'print(n = ...)' to see more rows
## [1] 0.66
```

Chicco and Jurman [14] compared *accuracy*, *mcc* and the F_1-measure for balanced and imbalanced datasets. They concluded that *mcc* is superior to the other performance measures for binary classifiers since it is unaffected by imbalanced datasets and produces a more informative and truthful performance measure. Delgado and Tibau [22] showed that *kappa* exhibits an undesirable behaviour in unbalanced situations, i.e., a worse classifier gets a higher value for *kappa*, unlike what is seen for *mcc* in the same situation.

For the DT classifier, the values of these measures are positive but relatively low in contrast to the optimistic 0.848 value obtained for *accuracy*.

```
# kappa and mcc for DT Classifier

# Key Counts from DT_pv_cm
tp <- 57; fn <- 22; fp <- 12; tn <- 132

# Cohen's kappa
A <- (tp + tn) / (tp + fn + fp + tn)
E <- ((tp + fp)*(tp + fn) + (fn + tn)*(fp + tn)) /
  (tp + fn + fp + tn)^2
(A - E) / (1 - E)
## [1] 0.657

# Matthews Correlation Coefficient
(tp*tn - fp*fn) / sqrt((tp + fp)*(tp + fn)*(fn + tn)*(fp + tn))
## [1] 0.66
```

You can, of course, use functions in the **yardstick** package to calculate these overall measures. The required commands are given below.

```
# Calculating kappa and mcc Using yardstick Functions

DT_pv %$% kap_vec(Survived, pred_class)
## [1] 0.657
DT_pv %$% mcc_vec(Survived, pred_class)
## [1] 0.66
```

2.3.3   Composite Measures

Threshold measures in this group are those obtained by combining selected class-specific measures. One such combination is *balanced accuracy* defined as the arithmetic average of *tpr* and *tnr* (taking the geometric mean yields a related measure called *G*-mean)

$$balanced\ accuracy = \frac{tpr + tnr}{2}. \tag{2.6}$$

In light of (2.3), this measure may interpreted as the value of *accuracy* that is obtained when the classification problem is perfectly balanced (i.e., when *prevalence* = 0.5). A related measure is the *J*-index defined by

$$J\text{-index} = tpr + tnr - 1 = 2 \times balanced\ accuracy - 1. \tag{2.7}$$

As noted earlier, common measures like *accuracy* are not suitable for problems with high class imbalance. Measures like *ppv* and *tpr* (i.e., *precision* and *recall*) and F_1-measure (also known as F_1-score) defined by their harmonic mean

$$F_1\text{-measure} = 2 \left[\frac{1}{ppv} + \frac{1}{tpr} \right]^{-1} = 2 \, \frac{ppv \times tpr}{ppv + tpr} \qquad (2.8)$$

are often more relevant (taking the geometric mean yields a related measure called *G*-measure). It is often a measure of choice when the focus in a binary classification problem is on the *positive* class. This measure allows you to make a trade-off between a classifier's ability to identify cases in the *positive* class and its ability to deliver correct *positive* predictions.

For example, with information retrieval systems, the goal is to retrieve as many relevant items as possible and as few nonrelevant items as possible in response to a request [93]. This requires use of classifiers that have high *recall* and *precision* (hence, F_1-measure) because you want the classifier to identify most of the relevant items and do it with high degree of accuracy.

Since (2.8) is focused on the *positive* class, we can also view it as a class-specific performance measure. Also, note that it may be expressed as [46]

$$F_1\text{-measure} = p \times recall + (1 - p) \times precision,$$

where $p = (tp+fn)/(2tp+fn+fp)$; the expression for p was given in Hand and Christen [46] using different notation. This points to one significant conceptual weakness of the F_1-measure, namely, the fact that the relative importance assigned to *precision* and *recall* is classifier dependent. Hand and Christen [46] highlighted this issue when evaluating classification algorithms for database record linkage and suggested that the same p should be used for all methods in order to make a fair comparison of the algorithms; see their article for how this can be accomplished.

One way to obtain the composite measures for the DT classifier is through the definitions for these measures. This, of course, is the obvious approach. It is a useful one to take initially since it reminds you of how the measures are defined.

```
# Composite Measures for the DT Classifier

# Key Counts from DT_pv_cm
tp <- 57; fn <- 22; fp <- 12; tn <- 132

# Class-Specific Measures
tpr <- tp / (tp + fn)
tnr <- tn / (fp + tn)
ppv <-  tp / (tp + fp)

# balanced accuracy
(tpr + tnr) / 2
## [1] 0.819
```

```
# Composite Measures for the DT Classifier (cont'd)

# J-index
tpr + tnr -1
## [1] 0.638

# F1-measure
2*(ppv * tpr) / (ppv + tpr)
## [1] 0.77
```

Alternatively, you can use **yardstick** functions to obtain the measures.

```
# Composite Measures Using yardstick Functions

DT_pv %$% bal_accuracy_vec(Survived, pred_class)
## [1] 0.819
DT_pv %$% j_index_vec(Survived, pred_class)
## [1] 0.638
DT_pv %$% f_meas_vec(Survived, pred_class)
## [1] 0.77
```

2.3.4   Interpreting Performance Measures

There are a number of issues to consider when you attempt to interpret the values you obtain for classifier performance measures. In this section, we examine some of these issues for threshold measures.

The first issue of interest is the range of possible values for each measure. For several measures, this range may be easily determined from their definitions (e.g., some like *accuracy* have an obvious range since they are fractional). Another issue is concerned with the question of what values to expect for useful/useless classifiers. Again, you can address this by careful examination of the definitions for some measures.

To get a feel for these concerns, let us start by examining the performance of three classifiers with key count vectors given in Table 1.3. What values do you expect to get for these classifiers when you compute the threshold performance measures? It is not difficult to answer this question by using the expressions for the key count vectors given in that table. We showed some results in Table 1.5. We can also proceed by considering a numerical example. Figure 2.9 shows the corresponding confusion matrices for the case when $m = n = 100$ and $\theta = 0.5$.

Next, we obtain the performance measures from the confusion matrices in Figure 2.9 including that given in Table 1.2 for the LM classifier in the first chapter.

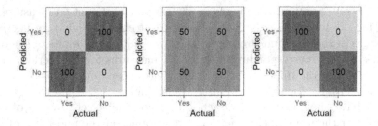

FIGURE 2.9
Confusion Matrices for Totally Useless, Random, and Perfect Classifiers

```
# LM versus Totally Useless, Random and Perfect Classifiers

# Function to Compute Performance Measures
measures <- . %>% con_mat(., c("Yes", "No")) %>% summary()

# Tabulate the Measures for Four Classifiers
list(c(0, 100, 100, 0), c(50, 50, 50, 50), c(100, 0, 0, 100),
    c(118, 32, 14, 136)) %>%
  map(~ measures(.) %>% select(-.estimator)) %>%
  reduce(inner_join, by = ".metric") %>%
  slice(1,2,7, 3,4,8, 11:13) %>%
  set_names(c("Measure", "TUC","RC","PC", "LM"))
## # A tibble: 9 x 5
##    Measure      TUC     RC     PC    LM
##    <chr>      <dbl>  <dbl>  <dbl> <dbl>
## 1 accuracy        0    0.5      1 0.847
## 2 kap            -1      0      1 0.693
## 3 mcc            -1      0      1 0.698
## 4 sens            0    0.5      1 0.787
## 5 spec            0    0.5      1 0.907
## 6 j_index        -1      0      1 0.693
## 7 precision       0    0.5      1 0.894
## 8 recall          0    0.5      1 0.787
## 9 f_meas          0    0.5      1 0.837
```

Note that, when defining the function to obtain the performance measures
from a given key count vector, we did not use function() (this is usually
used to define **R** functions). Instead, we use the function composition syntax
from the **magrittr** package. For more information on this convenient syntax
for composing **R** functions, see Mailund [75, p. 79], for example. Note the
use of a period placeholder for the argument to the function. In this case, it
represents a key count vector. You can infer this since it is the first argument
to the con_mat() function.

The second and fourth columns in the above tibble contain measures obtained for the totally useless (TUC) and perfect classifier (PC), respectively. The values in these columns represent the lower and upper limits of the range of values for the corresponding metric. All measures for the perfect classifier are equal to the upper limit of 1. With the exception of *kappa*, *mcc* and *J*-index, measures for the totally useless classifier are all equal to 0 (the exceptions have the value -1 as lower limit). For the selected value of θ, all measures for the random classifier (RC) take values at the mid-point of their respective ranges (in general, the measures depend on the value of θ).

As a general rule, we would like our trained classifier to have values for these measures that are (significantly) greater than those for the random classifier and as close to 1 as possible. This requirement is partially satisfied by the LM classifier. As can be seen by the values for *kappa*, *mcc* and *J*-index, the closeness to 1 requirement falls somewhat short of the ideal. The *accuracy* value is relatively high for the classifier and is greater than the estimated "No Information Rate" (whose estimated value here is 0.646). On the other hand, the other two overall measures for the classifier suggest mediocre performance.

The class-specific measures provide additional insights. We have already commented on *sensitivity* and *specificity* of the LM classifier in the first chapter. The relatively high *precision* is off-setted by the mediocre *recall* value. We see this reflected in the F_1-measure. Although the classifier does a relatively poor job at classifying *positive* cases, the correctness of such classifications is quite high.

In general, the practical relevance of the different class-specific measures depends on the substantive problem. When deciding which to focus on, you need to take into account the relative cost of classification errors, i.e., false negatives versus false positives. You can also think of comparing the cost of false omissions versus that of false discoveries. For example, in spam classification, false discovery is more serious than false omission. Hence, for this application, *precision* takes precedence over *recall*. On the other hand, *sensitivity* is more important for classifications involved with disease detection since it is important to control occurrence of false negatives.

2.4   Estimation of Performance Parameters

The binary CPA measures that we have considered earlier are generally viewed as descriptive numerical summaries of classifier performance. Some authors also view them as estimates in the proper statistical sense of the word. For example, you see this in the documentation by Sing et al. [101] for the **ROCR** package that you can use to produce performance curves for classifiers. This, in turn, implies that there are certain estimands involved when performance measures are being estimated. As seen in their documentation, these estimands refer to certain probabilities. Other examples include regarding *Precision* and

Recall as conditional probabilities by Hand and Christen [46], and applying Bayes theorem to derive the following relationship by Kuhn and Johnson [68, p. 41]

$$PPV = \frac{Sensitivity \times Prevalence}{Sensitivity \times Prevalence + (1 - Specificity) \times (1 - Prevalence)}$$
(2.9)

and a similar expression for *NPV*. Their derivation relied on thinking of *Prevalence*, *Sensitivity*, and *Specificity* as performance parameters that are defined by the relevant probabilities in Table 2.3; see Weiss [115] for a somewhat similar table. We review some of these parameters, including their relationships and estimation in this section.

2.4.1   Performance Parameters

To begin, note that it is important to make a distinction between descriptive classifier performance measures and the corresponding parameters.[10] In this book, the former comprises empirical metrics like those encountered in earlier sections of this chapter, and the latter refers to certain unknown probabilities like those in Table 2.3. For example, *accuracy* defined by (1.10) may be viewed as a descriptive performance measure, and *Accuracy* defined by $P(\widehat{Y} = Y)$ is the corresponding performance parameter. As can be seen from the table, (1.10) may also be viewed as an estimate of this probability. Also, note that definitions of Y and $\widehat{Y}$ in the table follow that given in Section 1.1.2 (this includes other relevant assumptions made in that section).

The probabilities in the second column of Table 2.3 are the estimands for the performance parameters in the first column (these probabilities may be found in Sing et al. [101] in different notation). The corresponding estimators are given in terms of key counts TP, FN, FP, and TN (random versions of those in Table 1.1) that arise when a classifier is applied to a random sample of test cases.

To illustrate the utility of the probabilities in Table 2.3, consider how to obtain (2.9). Instead of using Bayes theorem directly, we apply the Laws of Probability to obtain

$$PPV = P(Y = 1 \mid \widehat{Y} = 1)$$
$$= \frac{P(Y = 1, \widehat{Y} = 1)}{P(\widehat{Y} = 1)}$$
$$= \frac{P(Y = 1, \widehat{Y} = 1)}{P(Y = 1, \widehat{Y} = 1) + P(Y = 2, \widehat{Y} = 1)}$$
$$= \frac{P(\widehat{Y} = 1 \mid Y = 1)P(Y = 1)}{P(\widehat{Y} = 1 \mid Y = 1)P(Y = 1) + P(\widehat{Y} = 1 \mid Y = 2)P(Y = 2)},$$

---

[10]For performance parameters, we capitalize (partially or fully) the names they share with the corresponding descriptive performance measures, e.g., *Sensitivity* (*TPR*) is an example of the former while *sensitivity* (*tpr*) is an example of the latter.

and

$$P(Y = 2) \;=\; 1 - P(Y = 1),$$
$$P(\widehat{Y} = 1 \mid Y = 2) \;=\; 1 - P(\widehat{Y} = 2 \mid Y = 2).$$

When you use information in Table 2.3 to substitute the probabilities by the corresponding performance parameters, you get the desired result for *PPV*. Similarly, by applying these rules to re-express $P(Y = 2 \mid \widehat{Y} = 2)$, you get a similar expression for *NPV*; see Kuhn and Johnson [68, p. 41] for the expression that you wind up with.

2.4.2   Point Estimation of Performance Parameters

The estimators in Table 2.3 are the maximum likelihood estimators of the corresponding performance parameters. To see this, consider the following joint probability mass function (*jpmf*) of $\widehat{Y}$ and Y:

$$P(\widehat{Y} = i, Y = j) = \begin{cases} pq, & i = 1, j = 1 \\ p(1 - q), & i = 2, j = 1 \\ (1 - p)(1 - r), & i = 1, j = 2 \\ (1 - p)r, & i = 2, j = 2 \end{cases} \tag{2.10}$$

where p, q, and r are positive fractions. From this *jpmf*, it is easily follows that

$$P(Y = 1) = p, \; P(\widehat{Y} = 1 \mid Y = 1) = q \text{ and } P(\widehat{Y} = 2 \mid Y = 2) = r.$$

Note that p represents *Prevalence* and, in light of Table 2.3, it follows that q and r coincide with *True Positive Rate* and *True Negative Rate*, respectively. These three probabilities also determine all other performance parameters in Table 2.3. For example, you can re-express (2.9) as

$$PPV = \frac{qp}{qp + (1 - r)(1 - p)}.$$

As another example, note that

$$\begin{aligned} Accuracy &= P(\widehat{Y} = Y) \\ &= P(\widehat{Y} = 1, Y = 1) + P(\widehat{Y} = 2, Y = 2) \\ &= P(\widehat{Y} = 1 \mid Y = 1)P(Y = 1) + P(\widehat{Y} = 2 \mid Y = 2)P(Y = 2) \\ &= qp + r(1 - p). \end{aligned}$$

   To obtain the MLEs of the performance parameters in Table 2.3, let $(\widehat{Y}_1, Y_1), \ldots, (\widehat{Y}_n, Y_n)$ denote the predicted and actual responses that will be obtained when a trained binary classifier is applied to a random sample of n test cases. Also, let (TP, FN, FP, TN) denote the key count vector that

TABLE 2.3
Performance Parameters, Probabilities and Estimators

Performance Parameter	Probability	Estimator
Accuracy (A)	$P(\widehat{Y} = Y)$	$\frac{TP+TN}{TP+FN+FP+TN}$
Detection Prevalence (DP)	$P(\widehat{Y} = 1)$	$\frac{TP+FP}{TP+FN+FP+TN}$
DetectionRate (DR)	$P(Y = 1, \widehat{Y} = 1)$	$\frac{TP}{TP+FN+FP+TN}$
True Positive Rate (TPR)	$P(\widehat{Y} = 1 \mid Y = 1)$	$\frac{TP}{TP+FN}$
False Negative Rate (FNR)	$P(\widehat{Y} = 2 \mid Y = 1)$	$\frac{FN}{TP+FN}$
False Positive Rate (FPR)	$P(\widehat{Y} = 1 \mid Y = 2)$	$\frac{FP}{FP+TN}$
True Negative Rate (TNR)	$P(\widehat{Y} = 2 \mid Y = 2)$	$\frac{TN}{FP+TN}$
Positive Predicted Value (PPV)	$P(Y = 1 \mid \widehat{Y} = 1)$	$\frac{TP}{TP+FP}$
False Discovery Rate (FDR)	$P(Y = 2 \mid \widehat{Y} = 1)$	$\frac{FP}{TP+FP}$
False Omission Rate (FOR)	$P(Y = 1 \mid \widehat{Y} = 2)$	$\frac{FN}{FN+TN}$
Negative Predicted Value (NPV)	$P(Y = 2 \mid \widehat{Y} = 2)$	$\frac{TN}{FN+TN}$

results when predicted and actual responses are cross-tabulated. Given a re-alization (tp, fn, fp, tn) of this random vector, the likelihood function may be expressed as

$$L = constant \times (pq)^{tp} \times (p(1-q))^{fn} \times ((1-p)(1-r))^{fp} \times ((1-p)r)^{tn}.$$

On differentiating the log-likelihood, we obtain the following gradient vector

$$\begin{bmatrix} \frac{\partial \ln L}{\partial p} \\[2mm] \frac{\partial \ln L}{\partial q} \\[2mm] \frac{\partial \ln L}{\partial r} \end{bmatrix} = \begin{bmatrix} \frac{fn+tp}{p} - \frac{fp+tn}{1-p} \\[2mm] \frac{tp}{q} - \frac{fn}{1-q} \\[2mm] \frac{tn}{r} - \frac{fp}{1-r} \end{bmatrix}. \tag{2.11}$$

Next, set the right-hand side of (2.11) equal to the 3×1 zero vector and solve the resulting system of equations, to obtain

$$\hat{p} = \frac{tp + fn}{tp + fn + fp + tn}, \hat{q} = \frac{tp}{tp + fn} \text{ and } \hat{r} = \frac{tn}{fp + tn}.$$

The Hessian matrix of $\ln L$ is

$$\begin{bmatrix} -\frac{fn+tp}{p^2} - \frac{fp+tn}{(1-p)^2} & 0 & 0 \\[2mm] 0 & -\frac{tp}{q^2} - \frac{fn}{(1-q)^2} & 0 \\[2mm] 0 & 0 & -\frac{tn}{r^2} - \frac{fp}{(1-r)^2} \end{bmatrix}$$

and at $(\hat{p}, \hat{q}, \hat{r})$, this simplifies to

$$
\begin{bmatrix}
-\frac{n}{\hat{p}(1-\hat{p})} & 0 & 0 \\
0 & -\frac{n\hat{p}}{\hat{q}(1-\hat{q})} & 0 \\
0 & 0 & -\frac{n(1-\hat{p})}{\hat{r}(1-\hat{r})}
\end{bmatrix}.
$$

This matrix is negative definite if $\hat{p}$, $\hat{q}$, and $\hat{r}$ are all positive fractions. Under this condition, the solution given above for the likelihood equations are the maximum likelihood (ML) estimates of p, q, and r (the ML estimators follow when you replace the key counts by the corresponding random variables). Clearly, these estimates coincide with descriptive summaries *prevalence*, *tpr*, and *tnr*, respectively, that we considered earlier.

By the invariance property of MLEs, you can get the MLEs of the remaining estimands in Table 2.3 by substituting $\hat{p}$, $\hat{q}$ and $\hat{r}$ where appropriate. As seen in Table 2.4, the resulting MLEs coincide with the estimators in Table 2.3. For example, after making the required substitutions and simplifying, the MLE of *Accuracy* is

$$
(1 - \hat{p})\hat{r} + \hat{p}\hat{q} = \frac{TN + TP}{TN + FP + FN + TP}.
$$

Finally, note that the expected value of the negative Hessian matrix is equal to $n\boldsymbol{I}(p, q, r)$ where

$$
\boldsymbol{I}(p, q, r) =
\begin{bmatrix}
\frac{1}{p(1-p)} & 0 & 0 \\
0 & \frac{p}{q(1-q)} & 0 \\
0 & 0 & \frac{(1-p)}{r(1-r)}
\end{bmatrix}
$$

is the single observation Fisher information matrix. This matrix determines the asymptotic covariance matrix of the MLEs. It will used in the next section to construct large sample confidence intervals for TPR and TNR.

The MLEs of q and r (i.e., TPR and TNR) for the DT classifier are shown in the next code segment. As expected, the estimates coincide with the values of the corresponding performance measures given earlier.

```
# MLE of Performance Parameters for the DT Classifier

# Key Counts
tp <- 57; fn <- 22; fp <- 12; tn <- 132

# MLE of p, q and r
(p <- (tp + fn) / (tp + fn + fp + tn)) # Prevalence
## [1] 0.354
(q <- tp / (tp + fn)) # TPR
```

TABLE 2.4

MLEs of Binary Performance Parameters

Performance Metric	Estimand	MLE
Accuracy	$pq + (1-p)r$	$\frac{TP+TN}{TP+FN+FP+TN}$
Detection Prevalence (DP)	$(1-p)(1-r) + pq$	$\frac{TP+FP}{TP+FN+FP+TN}$
Detection Rate (DR)	pq	$\frac{TP}{TP+FN+FP+TN}$
True Positive Rate (TPR)	q	$\frac{TP}{TP+FN}$
False Negative Rate (FNR)	$1-q$	$\frac{FN}{TP+FN}$
False Positive Rate (FPR)	$1-r$	$\frac{FP}{FP+TN}$
True Negative Rate (TNR)	r	$\frac{TN}{FP+TN}$
Positive Predicted Value (PPV)	$\frac{pq}{(1-p)(1-r)+pq}$	$\frac{TP}{TP+FP}$
False Discovery Rate (FDR)	$\frac{(1-p)(1-r)}{(1-p)(1-r)+pq}$	$\frac{FP}{TP+FP}$
False Omission Rate (FOR)	$\frac{p(1-q)}{(1-p)r+p(1-q)}$	$\frac{FN}{FN+TN}$
Negative Predicted Value (NPV)	$\frac{(1-p)r}{(1-p)r+p(1-q)}$	$\frac{TN}{FN+TN}$

```
## [1] 0.722
(r <- tn / (fp + tn)) # TNR
## [1] 0.917


# MLE of Other Key Performance Parameters
p*q + (1-p)*r # Accuracy
## [1] 0.848
(p*q) / ((1-p)*(1-r) + p*q) # PPV
## [1] 0.826
(1-p)*r / ((1-p)*r + p*(1-q)) # NPV
## [1] 0.857
```

You can obtain the maximum likelihood estimates of other key parameters like *Accuracy*, *PPV*, and *NPV* in two ways. The first approach is to use the relevant formulas in the third column of Table 2.4, i.e., use the formulas we have seen before. Here, we used the second approach which makes use of the formulas in the second column and the MLEs of p, q, and r.

2.5 Other Inferences

Other inferences that you can make on performance parameters include interval estimation and tests of hypotheses. In this section, these alternative techniques for some key parameters will be discussed. To obtain the required inferences for the DT classifier, the relevant information is contained in the DT_pv tibble that contains the actual and predicted classes that were obtained when the classifier was applied to the test dataset; see Section 2.2.3.

2.5.1 Interval Estimation of Performance Parameters

To begin, let $\theta = P(\widehat{Y} = Y)$, i.e., θ is equal to the *Accuracy* parameter in Table 2.3, and consider finding a $100(1 - \alpha)\%$ confidence interval for it. To solve this problem, you can apply the exact approach in Mood et al. [81, p. 389]. The limits of the required interval (θ_L, θ_U) may be found by solving the following equations

$$\sum_{x=x_0}^{n} \binom{n}{x} \theta_L^x (1 - \theta_L)^{n-x} = \frac{\alpha}{2},$$

$$\sum_{x=0}^{x_0} \binom{n}{x} \theta_U^x (1 - \theta_U)^{n-x} = \frac{\alpha}{2},$$

where, in terms of key counts, $x_0 = tp + tn$ and $n = tp + fn + fp + tn$. The above equations may expressed as

$$\int_0^{\theta_L} \frac{1}{B(x_0, n - x_0 + 1)} v^{x_0-1} (1 - v)^{n-x_0} \, dv, = \frac{\alpha}{2},$$

$$\int_0^{\theta_U} \frac{1}{B(x_0 + 1, n - x_0)} v^{x_0} (1 - v)^{n-x_0-1} \, dv, = 1 - \frac{\alpha}{2},$$

since

$$\sum_{x=u}^{n} \binom{n}{x} \theta^x (1 - \theta)^{n-x} = \int_0^{\theta} \frac{1}{B(u, n - u + 1)} v^{u-1} (1 - v)^{n-u} \, dv, \qquad (2.12)$$

for positive integers u and n and $0 < \theta < 1$. This result is a re-expression of that given in Olkin et al. [86, p. 461] for the relationship between cumulative probabilities of the beta and binomial distributions.

Hence, the limits of the required confidence interval may be obtained by finding the relevant quantiles from the appropriate beta distributions, i.e.,

$$(\theta_L, \theta_U) = \left(Q_B \left(\frac{\alpha}{2}; x_0, n - x_0 + 1 \right), Q_B \left(1 - \frac{\alpha}{2}; x_0 + 1, n - x_0 \right) \right),$$

where $Q_B(p; a, b)$ is the p-th quantile of a $Beta(a, b)$ distribution.

The above confidence interval may be found using the acc_ci() function given in Appendix A.3. Using data in DT_pv, we obtain the 95% confidence interval for the *Accuracy* parameter given below.

```
# Estimating Accuracy for the DT Classifier

# Point Estimate & 95% Confidence Interval for Accuracy
DT_pv %$% table(pred_class, Survived) %>% acc_ci()
## # A tibble: 1 x 3
##    estimate lower_limit upper_limit
##       <dbl>       <dbl>       <dbl>
## 1    0.848       0.794       0.892
```

Note that the above confidence interval is one of the results you get when you apply the `confusionMatrix()` function from the **caret** package to obtain performance measures and related results for the DT classifier.

Given the key role played by TPR and TNR in binary CPA, it is also of interest to obtain confidence intervals for them. Recall, these parameters correspond to the q and r parameters, respectively, in the joint distribution given by (2.10) for the predicted target $\widehat{Y}$ and the actual target Y.

Using results given in the Section 2.4.2 on the MLEs of these parameters, you can obtain approximate confidence intervals. The limits of a large sample $100(1 - \alpha)\%$ interval for q are given by

$$\hat{q} \pm z_{\alpha/2} \sqrt{\frac{\hat{q}(1 - \hat{q})}{n\hat{p}}},$$

where $\hat{q}$ is th MLE of q, $z_{\alpha/2}$ is the $(1 - \alpha/2)$-th quantile of the standard normal distribution, n is the size of the dataset, and $\hat{p}$ is the MLE of p. The corresponding confidence limits for r are given by

$$\hat{r} \pm z_{\alpha/2} \sqrt{\frac{\hat{r}(1 - \hat{r})}{n(1 - \hat{p})}},$$

where $\hat{r}$ is th MLE of r. The two large sample confidence intervals follow from asymptotic properties of MLEs, and the general approach based on pivotal quantities for constructing confidence intervals that is discussed in Mood et al. [81, p. 387].

```
# Approx 95% CIs for the DT Classifier

# Sample Size and MLEs
n <- 223 # number of cases in DT_pv
p <- 0.354 # MLE of Prevalence
q <- 0.722 # MLE of TPR
r <- 0.917 # MLE of TPR
alpha <- 0.05; z_halfalpha <- qnorm(1 - alpha/2)
```

```
# 95% Confidence Interval for TPR
q + c(-1, 1) * z_halfalpha * sqrt(q*(1-q)/(n*p))
## [1] 0.623 0.821

# 95% Confidence Interval for TNR
r + c(-1, 1) * z_halfalpha * sqrt(r*(1-r)/(n*(1-p)))
## [1] 0.872 0.962
```

2.5.2   Testing Hypotheses on Performance Parameters

Despite some issues with *accuracy* [91], it remains a widely used performance measure in CPA. When it is applicable, it might also be of interest to know whether the classifier under evaluation has better accuracy compared to the classification rule that simply assigns every case to the majority class. This leads to the "No Information" hypothesis testing problem

$$H_0 : \theta \leq NIR \text{ versus } H_1 : \theta > NIR, \tag{2.13}$$

where $\theta = P(\widehat{Y} = Y)$ and NIR is a specified "No Information Rate". The latter is the value that one expects *a priori* for the population proportion of cases in the majority class. In practice, NIR is replaced by a suitable estimate. A classifier has no practical value if H_0 is not rejected for it.

A test of (2.13) may be performed with help from a one-sided lower $100(1 - \alpha)\%$ confidence interval for θ. Using arguments similar to what was done in Section 2.5.1, it can be shown that the required interval is given by $(Q_B(\alpha; x_0, n - x_0 + 1), 1)$ where x_0 is the observed value of the sum X of TP and TN. Thus, you can reject H_0 in (2.13) at $100\alpha\%$ level of significance if $NIR < Q_B(\alpha; x_0, n - x_0 + 1)$. Alternatively, you can use the following exact P-value

$$P(X \geq x_0) = \sum_{u=x_0}^{n} \binom{n}{u} p^u (1 - p)^{n-u} \text{ where } p = NIR,$$

$$= \int_0^{NIR} \frac{1}{B(x_0, n - x_0 + 1)} v^{x_0 - 1} (1 - v)^{n - x_0} \, dv$$

to perform the test. The integral on the right-hand side of the above expression follows when you apply (2.12).

For the DT classifier, the null hypothesis in (2.13) may be rejected given the P-value for the "No Information" test shown below.

```
# No Information Test P-Value for the DT Classifier

n <- 223; NIR <- 0.646
tp <- 57; tn <- 132; x0 <- tp + tn
pbinom(x0, n, NIR, lower.tail = FALSE) # P-value
## [1] 4.96e-12
```

Next, consider lower bounds q_0 and r_0 on TPR and TNR, respectively, for a given classification problem. In terms of parameters in the joint distribution given by (2.10), you can consider testing

$$H_0 : q \leq q_0 \text{ versus } H_1 : q > q_0$$

or

$$H_0 : r \leq r_0 \text{ versus } H_1 : r > r_0$$

separately, depending on whether your focus is on TPR or TNR.

As usual, you can solve the above hypothesis testing problems by computing the relevant P-value. Here, we take an interval estimation approach to solving them. To do this, consider the following approximate one-sided $100(1 - \alpha)\%$ confidence intervals

$$\left(\hat{q} - z_\alpha \sqrt{\frac{\hat{q}(1 - \hat{q})}{n\hat{p}}}, 1 \right)$$

and

$$\left(\hat{r} - z_\alpha \sqrt{\frac{\hat{r}(1 - \hat{r})}{n(1 - \hat{p})}}, 1 \right),$$

for the first and second hypothesis testing problem, respectively, where $\hat{q}$ and $\hat{r}$ are the MLEs given in Section 2.4.2.

You can reject the first null hypothesis at $100\alpha\%$ level of significance if the one-sided interval for q is contained within the interval $(q_0, 1]$. Similarly, you can reject the second null hypothesis if $(r_0, 1]$ contains the one-sided interval for r.

To illustrate, consider the case when $q_0 = 0.5$. Results from the following code segment shows that you can reject the first null hypothesis at 5% level of significance. You can also reject the second null hypothesis for the case when $r_0 = 0.5$.

```
# 95% LCIs for the DT Classifier

# Sample Size and MLEs
n <-  223; p <- 0.354; q <- 0.722; r <- 0.917

# 0.95 Standard Normal Quantile
alpha <- 0.05; z_alpha <- qnorm(1 - alpha)

# 95% Lower Confidence Interval for TPR
c(q - z_alpha*sqrt((q*(1-q))/(n*p)), 1)
## [1] 0.639 1.000

# 95% Lower Confidence Interval for TNR
c(r - z_alpha*sqrt((r*(1-r))/(n*(1-p))), 1)
## [1] 0.879 1.000
```

When both performance measures are relevant, you can consider testing

$$H_0 : (q,r) \notin \Psi \text{ versus } H_1 : (q,r) \in \Psi, \tag{2.14}$$

where

$$\Psi = \{(q,r) : q_0 < q < 1, r_0 < r < 1\}.$$

A special case of this problem is

$$H_0 : \min(q,r) \leq c_0 \text{ versus } H_1 : \min(q,r) > c_0, \tag{2.15}$$

where c_0 is a sufficiently large specified value. For example, if $c_0 = 0.5$, the alternative hypothesis states that the classifier in question is doing better than the random classifier when TPR and TNR are the parameters of interest. This hypothesis testing problem provides another way to assess the quality of a classifier.

If $\hat{q}_0$ and $\hat{r}_0$ are the observed values of $\hat{q}$ and $\hat{r}$, it follows the P-value for the test of (2.14) is given by

$$\begin{aligned} P(\hat{q} \geq \hat{q}_0, \hat{r} \geq \hat{r}_0) &= P(\hat{q} \geq \hat{q}_0) \times P(\hat{r} \geq \hat{r}_0) \\ &\approx \Phi\left(\frac{c_0 - \hat{q}_0}{\sqrt{c_0(1-c_0)/(n\hat{p})}}\right) \Phi\left(\frac{c_0 - \hat{r}_0}{\sqrt{c_0(1-c_0)/(n(1-\hat{p}))}}\right), \end{aligned}$$

where, as usual, $\Phi(\cdot)$ is the standard normal CDF. The above approximate P-value follows from the asymptotic normality and independence of the MLEs.

For the DT classifier, the P-value of the test of (2.15) with $c_0 = 0.5$ shows that H_0 can be rejected at 5% level of significance.

```
# P-value of Test for H0: min(q, r) <= c0

# Sample Size and MLEs
n <-  223; p <- 0.354; q <- 0.722; r <- 0.917

# Compute the P-Value
c0 <- 0.5
pnorm((c0 - q)/sqrt((c0*(1-c0))/(n*p))) *
  pnorm((c0 - r)/sqrt((c0*(1-c0))/(n*(1-p)))))
## [1] 2.75e-28
```

For binary classification problems, you can also use the test proposed by McNemar [79] to test the hypothesis of marginal homogeneity, i.e., the marginal distribution of actual and predicted response are the same. Since, we are dealing with binary responses, it suffices to state the null hypothesis as

$$H_0 : P(\hat{Y} = 1) = P(Y = 1). \tag{2.16}$$

This null hypothesis is equivalent to stating that the chance of getting a false positive is equal to that for getting a false negative. To see this, note that H_0 may be expressed as

$$P(\widehat{Y} = 1, Y = 1) + P(\widehat{Y} = 1, Y = 2) = P(\widehat{Y} = 1, Y = 1) + P(\widehat{Y} = 2, Y = 1).$$

After cancelling terms, it follows that (2.16) is equivalent to

$$H_0 : P(\widehat{Y} = 1, Y = 2) = P(\widehat{Y} = 2, Y = 1).$$

The P-value for testing this null hypothesis versus

$$H_1 : P(\widehat{Y} = 1, Y = 2) \neq P(\widehat{Y} = 2, Y = 1)$$

is given by $P(T \geq t_0)$ where t_0 is the observed value of

$$T = \frac{(FP - FN)^2}{FP + FN},$$

and $T \sim \chi_1^2$ provided $FP + FN$ is sufficiently large (i.e., at least 25).

Marginal homogeneity of predicted and actual responses is one of several criteria that can be used to assess the quality of a classifier. Its presence provides one indication of good performance by the classifier.

In light of the P-value computed below, the null hypothesis of homogeneity for the marginal distribution of the actual and predicted response is not rejected for the DT classifier at the usual 5% level of significance. Thus, for this classifier, the chance of getting a false positive is equal to that for getting a false negative.

```
# P-Value of McNemar's Test for DT Classifier

# Number of False Positives and False Negatives
fp <- 12 ; fn <- 22

# Observed Value of McNemar's Test Statistic
t0 <- (fp - fn)^2/(fp + fn)

# P-Value of McNemar's Test
pchisq(t0, 1, lower.tail = FALSE) # P-value
## [1] 0.0863
```

Other statistical tests have been used in CPA. A number of these arise when the problem involves comparative analysis of competing classifiers and in multiclass CPA. We'll mention some of these in the fourth and fifth chapters.

2.6   Exercises

1. The following confusion matrix from Zumel and Mount [119, p. 176] was obtained for a logit model classifier in a spam classification problem.

```
                prediction
truth        FALSE TRUE
  non-spam     264    14
  spam          22   158
```

   (a) Show how to put the given confusion matrix in the format shown in Table 1.1.

   (b) What is the *precision* of the logit model classifier under evaluation?

   (c) Of the emails that were non-spam, what fraction was incorrectly classified?

2. Consider classifier A and B with key count vectors given below.

```
kcv_A # key count vector for classifier A
##   tp  fn  fp  tn
## 500    0 200 300


kcv_B # key count vector for classifier B
##   tp  fn  fp  tn
## 300 200    0 500
```

   The counts are from two confusion matrices that were given in Provost and Fawcett [90, p. 191]. The authors used these array summaries to illustrate issues with the *accuracy* measure when class imbalance is significant.

   (a) Obtain the confusion matrices and take note of *accuracy* and *prevalence* for the two classifiers.

   (b) Compute *tpr* and *tnr* for the two classifiers.

   (c) Compare the *accuracy* of the two classifiers if *prevalence* is 0.1 instead of 0.5. Assume the same true rates as those obtained in part (b) for the classifiers.

   (d) Comment on the results obtained.

3. You can use the `crosstabToCases()` function in the **vtree** package to obtain the actual and predicted classes when given a confusion matrix as a "table" object. Part of the code for the `con_mat()` function in Appendix A.3 makes use of this function.

    (a) Use the above suggestion to obtain a tibble (containing the actual and predicted classes) from the confusion matrix in the answer to part (a) of Exercise 1. Check to ensure that both factors in the tibble have the *positive* class as the first level.

    (b) Show how to obtain the confusion matrix as a "conf_mat" object from the tibble in part (a). Hence, obtain a heat map of the resulting array summary.

    (c) Use functions from the **yardstick** package to compute *precision* and *recall* directly from information in the tibble of part (a).

4. The key counts $tp = 31$, $fn = 9$, $fp = 5$ and $tn = 30$ are from the confusion matrix reported in Nwangang and Chapple [83, p. 238] for a k-nearest neighbors classifier used in a binary heart disease classification problem.

    (a) Write an **R** function to compute *accuracy*, *kappa*, *mcc* and the F_1-measure from the above key counts.

    (b) Use your function and the above key counts to compute the measures mentioned in part (a). Show how to check your answers with help from the **yardstick** package.

5. (a) Make a qualitative comparison of the relative cost of false positive and false negative errors that you can make when you classify emails in a binary spam classification problem.

    (b) Repeat part (a) when you classify individuals in a binary default detection problem.

    (c) Write an **R** function that will enable you to compute a binary F_β-score. Apply your function using information on the two confusion matrices given in Figure 1.3 of Chapter 1. For choice of β, see the flowchart in Brownlee [10] (the measure is defined on p. 40 of this reference).

6. Table 4 in Chicco and Jurman [14, p. 9] contain the four key counts from six confusion matrices. The table also contain results comparing *accuracy*, *mcc* and the F_1-measure for balanced and imbalanced datasets. The results provide support for the authors argument that *mcc* is superior to the other two performance measures for binary classifiers.

    (a) Show how to obtain heat maps of the six confusion matrices.

    (b) Obtain an estimate of *prevalence* from each confusion matrix.

    (c) Obtain estimates of *accuracy*, *kappa*, *mcc*, and F_1-measure. Comment on the results.

7. One measure you can use to distinguish between *positive* and *negative* cases is *discriminant power* (*dp*). Sokolova et al. [102] showed

that this measure may be expressed as

$$dp = \frac{\sqrt{3}}{\pi} \ln \left(\frac{\rho_+}{\rho_-} \right),$$

where

$$\rho_+ = \frac{sensitivity}{1 - specificity} \text{ and } \rho_- = \frac{1 - sensitivity}{specificity}$$

refer to composite performance measures called *positive likelihood* and *negative likelihood*, respectively.

(a) Write an **R** function to compute *dp*. Assume the input to your function is a vector of key counts from a binary confusion matrix. Allow your function to return either *dp* or a list containing this value and the likelihoods.

(b) Apply your function to compute *dp* for the decision tree classifier that we covered in the text; see Figure 2.5 for the confusion matrix. Interpret the value you obtain. Note that some guidance is available in Sokolova et al. [102].

8. The three confusion matrices in Fernandez et al. [36, p. 50] are not in our preferred format; see Table 1.1.

(a) Identify the corresponding 1×4 key count vectors. Use the same format as that in (1.9).

(b) Construct the corresponding confusion matrices in our preferred format.

(c) Obtain the corresponding overall threshold measures.

(d) Comment on your results.

3

Classifier Performance Curves

The performance measures discussed in the previous chapter have one serious disadvantage for an important category of classifiers. For binary classification problems, scoring classifiers rely on choice of an appropriate threshold to decide whether a case belongs to the *positive* class. Threshold performance measures like *accuracy* and class-specific ones like *sensitivity* and *specificity* depend on choice of the threshold. To see this, it suffices to note that for classifiers with classification rules like that given by (1.2) in Chapter 1, the number of *positive* classifications decreases as we increase the threshold t.

Performance curves like those that we will discuss in this chapter provide a way to deal with the abovementioned issue. Another advantage of such curves is the fact that they not only give you a visualization of classifier performance, but they also provide informative scalar metrics that cover aspects of performance not included in threshold measures. The ROC curve of a scoring classifier is a prime example of performance curves. Figure 1.4 in the first chapter is one example of such a curve.

In this chapter, we provide further details on ROC curves and performance measures that you can derive from them. We also discuss Precision-Recall (PR) curves that may be more useful in some applications. We omit discussion of other performance curves like profit and lift curves. Interested readers may find some information on these curves in Provost and Fawcett [90], for example.

3.1 Another Classification Machine Learner

Neural networks provide a useful alternative approach to solve supervised (and unsupervised) learning problems; see Roiger and Geatz [97, p. 256] for the strengths and weaknesses of this machine learning technique. They have been widely applied in business [111] and medicine [1, 31], for example. For a comprehensive survey of other applications, see Widrow et al. [117].

In this section, we train a neural network (NN) classifier for the Titanic survival classification problem that we considered in Section 2.2. The resulting trained classifier will be partially evaluated when we use it to illustrate the various performance curves and measures that will be introduced later in this chapter. Its evaluation will be completed in the next chapter when we compare its performance with that of three other classifiers for the same classification problem.

3.1.1   Training the NN Classifier

A neural network attempts to mimic the way neurons in a human brain communicate with one another to arrive at a decision. Several architectures are available for such a machine learner; the relevant one depends on the problem you want to solve. For classification problems, the one to use is a fully connected feed-forward neural network like that shown in Figure 3.1. This figure is a display of the trained neural network (NN) classifier for the problem in Section 2.2. It is based on a preprocessed version of the `Titanic_train` dataset from Section 2.2.2; the preprocessing involves use of the `step_range()` function to transform the numeric features so that they range over the $[0, 1]$ interval; this function does the same thing as the min-max normalization function in Nwanganga and Chapple [83, p. 245].

```
# Training the NN Classifier

# Obtain Preprocessed Training Data
library(recipes) # core package in tidymodels
Train_data <-
  recipe(Survived ~ ., data =  Titanic_train) %>%
  step_range(Age, FamSize) %>%
  prep() %>%
  juice()

# Train the NN Classifier
library(nnet) # help(package = "nnet")
set.seed(280623) # for nnet initial weights
NN_fit <- mlp(mode = "classification", engine = "nnet") %>%
  fit(Survived ~., data = Train_data)
```

In the above code segment, we obtain `Train_data` by (i) using `recipe()` and `step_range()` to create a `"recipe"` object, i.e., a data structure that represents the required preprocessing pipeline, (ii) applying `prep()` to this object to obtain any estimates required for the preprocessing (here, this means finding minimum and maximum values), and (iii) applying `juice()` to the result to extract the preprocessed training data. All the abovementioned functions are from the **recipes** package.

Training of the classifier was done using the `mlp()` function from the **nnet** package. To display the fitted NN classifier, we used the `plotnet()` function from the **NeuralNetTools** package.

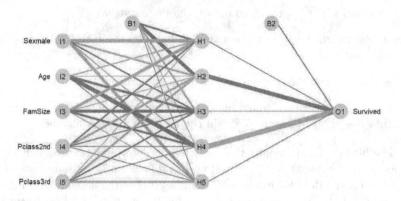

FIGURE 3.1
Neural Network Classifier for Titanic Survival Classification

```
# Display the Trained NN Classifier

library(NeuralNetTools) # help(package = "NeuralNetTools")
NN_fit %>% extract_fit_engine() %>%
  plotnet(cex_val = 0.65, circle_cex = 3, pos_col = "seagreen",
    neg_col = "indianred1")
```

The trained (i.e., fitted) NN classifier displayed in Figure 3.1 assigns a case to the *positive* class if the corresponding membership score is sufficiently large. Specifically, this assignment is made if

$$\phi_0 \left(\alpha + \sum_{h=1}^{5} \omega_h \phi_h \left(\beta_h + \sum_{i=1}^{5} w_{ih} x_i \right) \right) > t, \tag{3.1}$$

where t is a suitable threshold (with 0.5 as default value). The left-hand side of (3.1) is the value of the output node (the right-most node). This value is the *positive* class membership score that is assigned by the NN classifier to a passenger with features captured by the input nodes (the five left-most nodes). Here, it tells you how strong the evidence is for a passenger to be classified as a survivor (recall, survivors of the Titanic sinking belong to the *positive* class). The various terms on the left-hand side of (3.1) are as follows:

- x_i is the feature value received by the i-th input node,

- w_{ih} is the weight given to the connection between i-th input node and h-th hidden node (this is one of the five nodes in the middle layer),

- β_h is the bias attached to the h-th hidden node,

- $\phi_h(\cdot)$ is the activation function for the h-th hidden node that is used to transform $\beta_h + \sum_{i=1}^{5} w_{ih}x_i$ to obtain the node output,

- ω_h is the weight given to the connection between the h-th hidden node and the output node,

- α is the bias attached to the output node,

- $\phi_0(\cdot)$ is the activation function to transform the value of the expression within the outer brackets to obtain the score for the output node.

The bias terms and connection weights are determined by the backpropagation algorithm; see Chapter 8 in Roiger and Geatz [97] for an outline of this algorithm and a detailed example illustrating it. For the fitted neural network in Figure 3.1, the complete set of values for the weights and bias terms may be obtained by using the code below (output omitted).

```
# Print the Fitted Biases & Weights

NN_fit %>% extract_fit_engine() %>% summary()
```

3.1.2   Predictions from the NN Classifier

Since our focus in this chapter is on performance curves, it suffices to have predictions of *positive* class membership probabilities (or suitable scores that measure the strength of *positive* class membership) for cases in the test dataset, and information on the corresponding actual classes. However, given what we plan to do in the next chapter, we also obtain predicted classes of the test cases.

To begin, we need to preprocess `Titanic_test` to obtain `Test_data`. This may be done by using the `bake()` function from the **recipes** package. Basically, what you need to do is repeat the first two steps (i.e., (i) and (ii)) that was used to produce `Train_data`, and then follow up with an application of `bake()`. This allows application of the same preprocessing estimates from these two steps to preprocess `Titanic_test`.

```
# Predictions from the NN Classifier

# Obtain Preprocessed Test Data
Test_data <-
    recipe(Survived ~ ., data = Titanic_train) %>%
    step_range(Age, FamSize) %>%
    prep() %>%
    bake(Titanic_test)
```

```
# Predictions from the NN Classifier (cont'd)

# Obtain Predictions
NN_pv <-
  bind_cols(
    predict(NN_fit, new_data = Test_data, type = "prob"),
    predict(NN_fit, new_data = Test_data, type = "class"),
    Test_data %$% Survived
  ) %>%
  set_names(c("prob_No", "prob_Yes", "pred_class", "Survived"))

# Save Predictions in a CSV File
# NN_pv %>% write_csv("NN_pv.csv") # save to working directory

NN_pv %>% print(n = 3)
## # A tibble: 223 x 4
##    prob_No prob_Yes pred_class Survived
##      <dbl>    <dbl> <fct>      <fct>
## 1    0.464    0.536 Yes        Yes
## 2    0.686    0.314 No         No
## 3    0.269    0.731 Yes        No
## # ... with 220 more rows
```

The variables in `NN_pv` have the same labels as in the `DT_pv` tibble of the last chapter. The information we need in this chapter are contained in `prob_Yes` and `Survived`. Note that "Yes" is the second level for the two factor variables in `NN_pv`. Since we want it to be the first level, we make the required change for both factors.[1]

```
# Check Reference Levels
NN_pv %>% keep(is.factor) %>% map(~ levels(.))
## $pred_class
## [1] "No" "Yes"
##
## $Survived
## [1] "No" "Yes"

# Make Yes the Reference (i.e., Positive) Class
NN_pv <- NN_pv %>% mutate_if(is.factor, fct_rev)
```

---

[1]The change is also made for `pred_class` since we will use it in the next chapter to compare threshold performance measures for the NN classifier with others for the problem in Section 2.2.

3.2   Receiver Operating Characteristics (ROC)

Recall from Chapter 1 that the ROC of a classifier encapsulates information on how *tpr* varies with *fpr* as we vary the threshold of a scoring classifier. You can get a visualization of the variation in question through the ROC curve and obtain a summary measure like *AUC* from the curve. The plot and measure play an important role in applications that involve ROC analysis. For example, in diagnostic testing, such analysis is used to determine the ability of a test to discriminate between groups, choose the threshold for a test, and compare performance of two or more tests [87]. Recall, we earlier identified several other application areas for ROC analysis in the first chapter.

There are other aspects to ROC curves and the areas under them that are less well known, e.g., the idea that the curve and *AUC* help one to determine the ability of a classifier to produce good relative class membership scores. Also, rather than think of *AUC* as a geometric property of the ROC curve, it is sometimes useful to think of it in probabilistic terms. These aspects and others will be discussed in this section.

3.2.1   ROC Curves

As noted earlier, to obtain an ROC curve of a trained classifier from test data, you need information on predicted *positive* class membership probabilities and actual classes from the data. For the NN classifier, you can get the required information from the second and fourth column of the NN_pv tibble that was obtained in the last section. Given this information, you can obtain the key count vector $(tp(t), fn(t), fp(t), tn(t))$ for a given threshold t in (3.1), and hence the pair of rates

$$(fpr(t), tpr(t)) = \left(\frac{fp(t)}{fp(t) + tn(t)}, \frac{tp(t)}{tp(t) + fn(t)} \right).$$

By varying t and plotting the resulting pairs, you can obtain the ROC curve of the NN classifier. This is how you can proceed in the absence of a function like roc_curve() from the **yardstick** package. When we use this function, we obtain the performance curve shown in Figure 3.2.

```
# ROC Curve for the NN Classifier

library(yardstick)
NN_pv %>% roc_curve(Survived, prob_Yes) %>% autoplot()
```

The ROC curve of the NN classifier shows that it performs significantly better than one that assigns a passenger at random to the survivor class (i.e., the "Yes" class). The initial steepness of the curve is a plus point since

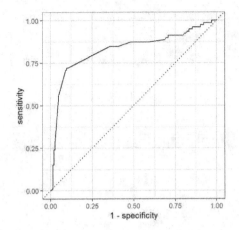

FIGURE 3.2
ROC Curve for the NN Classifier

benefits (i.e., high *tpr*) are rapidly gained with low cost (i.e., low *fpr*). This is a desirable quality. However, this is offsetted by the slowness at gaining maximal benefit as cost increases. In other words, the curve falls somewhat short of what is expected of the ideal ROC curve given in the first chapter. The shortfall in performance may be quantified by examining the *AUC*, i.e., area under the curve (we discuss this in the next section).

What else can you learn from the ROC curve? The answer is that the curve has something to say about the separability of scores (or class membership probabilities, when applicable). More precisely, it tells you something about the separation between the distribution of *positive* class membership scores (or probabilities) for cases in *positive* and *negative* classes. Good separation (in the right direction) is needed for good performance to be supported by the ROC curve (and by its *AUC*). Figures 3.3 and 3.4 demonstrate this point.

Figure 3.3 illustrates the different degrees of separation in the distributions of *positive* class membership scores for the two classes. Each ROC curve in Figure 3.4 is identified by the quality of separation. This quality refers to how well the distribution of *positive* class membership scores separate for the two classes in a binary classification problem.

Good separation is required in order for the ROC curve to lie in the upper left rectangle of the unit square. This is a basic requirement of a good classifier. It is important to note that the separation needs to be in the right direction. The plots in the top right panel of both figures show why this is so (this does not mean that the corresponding classifier is useless because you can modify the classification rule for it to function like a classifier with the ideal ROC curve). For classifiers like those represented by (1.2) in the first chapter, the distribution of *positive* class membership scores for the *positive* class should be to the right of the one for the *negative* class.

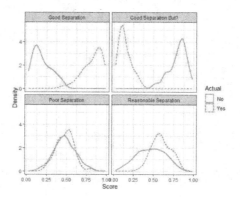

FIGURE 3.3
Separability of Positive Class Membership Score Distributions

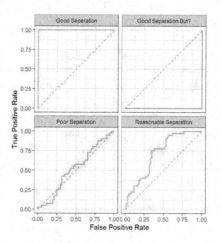

FIGURE 3.4
Impact of Difference in Separability on ROC Curves

Although ROC curves are widely used to evaluate classifiers in the presence of class imbalance, it can be optimistic for severely imbalanced class problems with few cases of the minority class [10, 19]. To overcome this problem, Drummond and Holte [26] suggested the use of cost curves.

3.2.2 Area Under the ROC Curve

The area under an ROC curve (AUC) is more than a geometric property of the curve. It is an example of a ranking measure according to the taxonomy introduced by Ferri et al. [37], and there are several alternative interpretations for it. You can view it as the average sensitivity if specificity values are chosen uniformly; for classification rules like (1.2) in Chapter 1, it is equivalent to the probability that a randomly chosen case from the *positive* class will yield a higher score than one randomly selected from the *negative* class [57]. Other interpretations may be found in Hand and Anagnostopoulos [54], for example.

The second alternative interpretation is related to separability of score distributions. The AUC for the curves in Figure 3.4 and some threshold measures are given in below.

```
# Performance Measures & Separability of Scores

## # A tibble: 4 x 5
##    plot                    AUC accuracy precision recall
##    <chr>                   <dbl>   <dbl>     <dbl>  <dbl>
## 1 Good Separation         1       0.99      1      0.975
## 2 Good Separation But?    0       0         0      0
## 3 Poor Separation         0.519   0.55      0.442  0.475
## 4 Reasonable Separation   0.713   0.64      0.534  0.775
```

There are several points to note from the above results. First, it shows that good separation in score distributions can result in $AUC = 1$ (the value one gets for the ideal ROC curve) even though the corresponding *accuracy* is less than ideal. Second, good separation but incorrect relative position of the score distributions yields a totally useless classifier if you do not re-define the *positive* class. Third, when there is poor separation in the distributions, the classifier functions like one that makes random classifications.

For the NN classifier, the AUC is equal to 0.829. You can infer from this that separability of the score distributions is more than reasonable for this classifier. The AUC value is significantly better than the 0.5 value for a random classifier. Following the interpretation guidelines given in Table 4 in Nahm [82], you can conclude that the value indicates good performance by the classifier.

```
# AUC for the NN Classifier

NN_pv %>% roc_auc(Survived, prob_Yes)
## # A tibble: 1 x 3
##   .metric .estimator .estimate
##   <chr>   <chr>          <dbl>
## 1 roc_auc binary         0.829
```

We used the `roc_auc()` function in the **yardstick** package to obtain AUC for the NN classifier. Without this functionality, you can obtain the required area by numerical integration or use a statistical approach. It is instructive to examine the second approach since it plays a role in a later discussion of the Hand and Till [56] estimate of AUC for a multiclass classifier. To illustrate the approach, we will use a dataset from Nahm [82].

To begin, consider the scoring classifier (1.2) in Chapter 1. As noted earlier, the area under the ROC curve (as a parameter) for such a classifier is equivalent to the probability that a randomly selected *positive* case will have a larger score than a randomly selected *negative* case, i.e.,

$$\text{AUC} \equiv P(S_{1|1} > S_{1|2}), \tag{3.2}$$

where $S_{1|c}$ is the class 1 membership score for a randomly selected case from class c. Here, think of class 1 as the *positive* class.

To estimate the right-hand side of (3.2), Hand and Till [56] suggested the estimator

$$\hat{A} = \frac{S - n_1(n_1 + 1)/2}{n_1 n_2}, \tag{3.3}$$

where S is the sum of ranks of the $S_{1|1}$ scores when these values are combined with the $S_{1|2}$ scores and the resulting set of $(n_1 + n_2)$ values are arranged in increasing order (the authors also provided a brief argument that links (3.3) to the AUC of a step function ROC curve). Note that scores in Hand and Till [56] refer to estimated class membership probabilities. The same is true for scores in Kleiman and Page [66]; see their paper for an alternative formula for the estimator of the right-hand side of (3.2).

To illustrate calculation of (3.3), consider using information in the small dataset given in Nahm [82, p. 29]; the author used the data to construct the ROC curve in his article. The simple step function form of the curve allows you to easily calculate the area under it by elementary geometric arguments. Thus, you can easily check correctness of the value that we obtain for $\hat{A}$ in the following code segment.[2]

---

[2]This check is left for the reader to do in Exercise 3; two other ways of getting the AUC are also included as part of the exercise.

```
# ROC AUC Using Data from Nahm (2021, p. 29)

## # A tibble: 10 x 3
##      case cancer marker
##     <int> <fct>   <dbl>
## 1      1 No       25.8
## 2      2 No       26.6
## 3      3 No       28.1
## 4      4 Yes      29
## 5      5 No       30.5
## 6      6 No       31
## 7      7 No       33.6
## 8      8 No       39.3
## 9      9 Yes      43.3
## 10    10 Yes      45.8

# Estimate AUC Using Formula (3) in Hand & Till (2001)
S <- 4 + 9 + 10; n1 <- 3; n2 <- 7
(S - n1*(n1+1)/2)/(n1*n2) # Ahat
## [1] 0.81
```

As a scalar descriptive measure of classifier performance, AUC is preferred to *accuracy* [7]. The utility of AUC is not affected by class priors, and its use does not require specification of misclassification costs and a classification threshold. These properties were noted by Hand and Till [56] among others.

Huang and Ling [61] showed that AUC is a statistically consistent and more discriminating measure than *accuracy*. Empirical comparisons between AUC and Matthews correlation coefficient (*mcc*) were made by Halimu et al. [45]. They found that both measures are statistically consistent with each other, and that AUC is more discriminating than *mcc*. However, in a recent study, Chicco and Jurman [15] presented arguments to support their proposal that *mcc* should replace the AUC as the standard metric for assessing binary classifiers. Conflicting results like these should come as no surprise given the fact that the measures being compared quantify different aspects of classifier performance.

Another noteworthy point is the fact that there are applications for which AUC is not even a contender (e.g., for spam detection, the F_1-measure is preferred since minimizing false discoveries is more important in this application), and there are also concerns about the incoherency of AUC and its use when comparing performance of competing classifiers. We take up this issue next.

While you can use AUC as a descriptive summary measure when looking at the ROC of a *selected* classifier (assuming selection has been done using other coherent measures of performance), its use in a comparative CPA of competing classifiers can be problematic. Hand [50] noted that when the curves cross, AUC can give potentially misleading results. He also noted a more serious deficiency with its use, namely, AUC implicitly uses different misclassification cost distributions for different classifiers (this is the reason for the incoherency). This amounts to saying that AUC uses different metrics to evaluate different classification rules.

To solve the abovementioned problems, Hand [50] proposed a coherent alternative measure

$$H = \frac{L_{ref} - L}{L_{ref}}, \tag{3.4}$$

where

$$L = \int_0^1 [c(1 - \pi)(1 - G(c)) + (1 - c)\pi F(c)]w(c)dc, \tag{3.5}$$

represents the overall expected minimum misclassification loss, and L_{ref} is the maximal value of L that is obtained when $F \equiv G$ (i.e., when the score distributions are not separated at all; this happens when you use a random classifier). On the right-hand side of (3.5), c represents the (normalized) cost of a false positive, π is the prevalence (i.e., prior) of the *positive* class, $F(\cdot)$ and $G(\cdot)$ are the cumulative distribution functions of the score for a case randomly selected from the *positive* and *negative* class, respectively, and $w(\cdot)$ represents the cost density function.

Note that (3.5) is essentially that given in Hand and Anagnostopoulos [53] with some changes in notation. The authors noted that this loss function is the result of "balancing costs over a distribution of likely values for misclassification costs". The measure defined by (3.4) was derived by Hand [50] with this perspective in mind and the assumption that cost of a false positive is a function of relative cost of misclassification, scores are fractional, and a case is classified as *positive* if the corresponding score is greater than the specified threshold.[3] Following his proposal, it was discovered that H-measure originated from work by Buja et al. [11] on the use of boosting methods to improve classification algorithms; see Hand and Anagnostopoulos [53] for some discussion of this alternative derivation for the measure. The alternative independent justification for H-measure reinforces the value of this coherent metric.

When applied to a classifier, the numerator of (3.4) represents the reduction in expected minimum misclassification loss when compared to L_{ref}, the corresponding maximal loss for a random classifier. H-measure expresses this reduction as a fraction of L_{ref}. Its range follows from its fractional nature; values close to 1 are preferred for a classifier.

Hand [50] showed that

$$AUC = 1 - \frac{L_d}{2\pi(1 - \pi)}, \tag{3.6}$$

---

[3] Henceforth, we'll use the term H-measure as in Hand and Anagnostopoulos [53] rather than H when referring to (3.4).

where L_d is what you get for (3.5) when you set the cost distribution $w(\cdot)$ equal to the mixture score density function

$$d(s) = (1 - \pi)g(s) + \pi f(s), \quad 0 < s < 1.$$

Here, $f(\cdot)$ and $g(\cdot)$ are density function corresponding to $F(\cdot)$ and $G(\cdot)$, respectively. It is important to note that, in general, density function $d(\cdot)$ is not the same for different classifiers [50, 53]. When used to obtain L_d, this means the result is obtained using a cost distribution that differs with different classifiers! This is a fundamental flaw that points to the incoherence of AUC highlighted by Hand [50] since the cost distribution is a property of the classification problem, not the classifier.

With sufficient domain knowledge, you can specify $w(\cdot)$ but, although this may be ideal, practical application of this approach can be problematic. Thus, it is useful to have a standard specification that produces the same summary measure from the same data [53]. Such a specification is given by the default beta distribution recommended by Hand and Anagnostopoulos [54]. In practice, their suggested $Beta(\alpha, \beta)$ distribution with density function

$$w(c) = \frac{c^{\alpha-1}(1-c)^{\beta-1}}{B(\alpha, \beta)}, \quad 0 < c < 1,$$

where $\alpha = 1 + \pi$ and $\beta = 2 - \pi$ is conditional on π because of the uncertainty in prevalence; this uncertainty may be modeled by a $Beta(2, 2)$ distribution. In any case, the same weight distribution should be used when comparing different classifiers for a given problem.

A Monte Carlo approach based on random variates from the two beta distributions mentioned earlier may be used to estimate H-measure; see Hand and Anagnostopoulos [53] for a revised expression of the measure and other relevant details involved in this approach. Fortunately, we are spared the effort since the HMeasure() function in the **hmeasure** package may be used to obtain the required estimate (for convenience, we use the wrapper function HM_fn() that is given below because we want only the value of the H-measure to be returned when a call is made to calculate it).

```
# H-Measure for the NN Classifier

library(hmeasure)
HM_fn <- function(grp, prob_Y){
  HMeasure(grp, prob_Y) %>% .$metrics %$% H
}
NN_pv %$% HM_fn(Survived, prob_Yes)
## [1] 0.483
```

The above calculation shows that the reduction in expected minimum misclassification loss for the NN classifier is about 48% of the corresponding loss for a random classifier. This improvement in performance is rather mediocre.

Readers who want further information on H-measure should refer to the excellent exposition of it in Hand and Anagnostopoulos [53]. Those interested in practical relevance of the measure may find Hand [51] an interesting alternative reference, and those who want a more in-depth discussion on the theoretical foundations of the measure should refer to the original article by Hand [50]. The second article highlights an interpretation of AUC that points to a fundamental weakness of this measure for assessing the effectiveness of diagnostic tests in medicine, and presents H-measure as a solution to the problem.[4]

H-measure is here to stay, like *mcc*, *sensitivity* and *specificity* (these threshold measures have a sound basis for their use in practice). It is already widely accepted in the machine learning and related communities, and indications are that its acceptability will continue to rise.

3.3   Precision-Recall (PR) Curves

Curves based on *precision* and *recall* are recommended for highly imbalanced domains where ROC curves may provide an excessively optimistic view of the performance [8]. Since such curves focus on the minority class, it is an effective diagnostic for imbalanced binary classification models [10]. The basic approach to construction of these curves is similar to that for the ROC curve.

Recall, the ROC curve is based on the variation of the pair of performance measures (fpr, tpr) as thresholds change for a scoring classifier. Measures like *precision* also change with thresholds. You get the Precision-Recall (PR) curve when you plot the collection of $(recall, precision)$ points as the threshold for the classifier is varied.

For example, the PR curve for the LM classifier given by (1.6) in Chapter 1 is displayed in Figure 3.5. Also shown in the figure is the ideal PR curve that is determined by points in $\{(0, 1), (1, 1), (1, 0.5)\}$. The point on this curve at the top right corner yields the ideal combination of *precision* and *recall*. The dotted horizontal line is the expected PR curve of a random classifier. This line is determined by the estimated *prevalence* of the *positive* class (this explains the second coordinate of the third point in the above set of points). A poor classifier will have a PR curve in the vicinity of the horizontal dotted line. For such a classifier, the area under the PR curve is determined by the estimated *prevalence*.

Also plotted in Figure 3.5 are points corresponding to $(recall, precision)$ pairs for three thresholds. Notice that *precision* (*recall*) increases (decreases) as the threshold for the classifier increases.

---

[4]When reading the suggested articles, take note that it helps to think of class 1 in the articles as the *positive* class, c as the normalized cost of a false *positive*, and a case is classified as *positive* if its classification score exceeds a threshold.

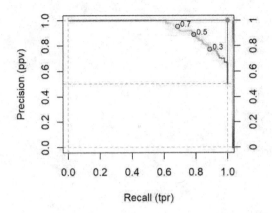

FIGURE 3.5
PR Curve for the LM Classifier in Chapter 1

```
# Three (recall, precision) Points on the PR Curve

## # A tibble: 3 x 4
##       c      t   recall precision
##    <dbl> <dbl>    <dbl>     <dbl>
## 1  0.3 -0.847    0.887     0.773
## 2  0.5  0        0.787     0.894
## 3  0.7  0.847    0.680     0.953
```

The PR curve for the neural network (NN) classifier is given in Figure 3.6. The *AUC* in this case is slightly more than twice the estimated *prevalence* but still falls somewhat short of the ideal value of 1.

```
# PR Curve and AUC for the NN Classifier

# Plot the PR Curve
NN_pv %>% pr_curve(Survived, prob_Yes) %>% autoplot()

# Estimate the AUC
NN_pv %>% pr_auc(Survived, prob_Yes) %>% pluck(".estimate")
## [1] 0.721

# Prevalence
NN_pv %$% table(Survived) %>% prop.table() %>% pluck("Yes")
## [1] 0.354
```

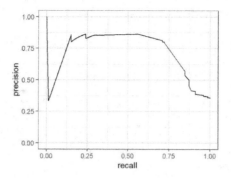

FIGURE 3.6
PR Curve for the NN Classifier

PR curves are more informative than ROC curves when dealing with highly imbalanced data. Saito and Rehmsmeier [99] noted that use of ROC curves for imbalanced problems can lead to deceptive conclusions about the reliability of classification performance; on the other hand, PR curves are more reliable because they involve the fraction of correct positive predictions (this is what *precision* quantifies). Davis and Goadrich [19] noted that looking at PR curves can expose differences between classification algorithms that are not apparent in ROC space; see Figure 1 in their paper for an illustration of this point. They also showed that a curve dominates in the ROC space if and only if it dominates in the PR space, and gave a counter example to show that an algorithm that optimizes the area under the ROC curve is not guaranteed to optimize the area under the PR curve.

3.4   Exercises

1. Consider the decision tree classifier displayed in Figure 2.3 for the Titanic survival classification problem. Use the predictions contained in the DT_pv.csv file for this exercise (the file is available in the publisher's website). This file contains the same information as in the DT_pv tibble that was obtained earlier when the classifier was applied to some test data.

    (a) Obtain density plots of *positive* class membership probabilities for the two classes.

    (b) Obtain the ROC curve and corresponding *AUC*.

    (c) Compute the *H*-measure.

2. This exercise is based on synthetic data in the following tibble. It illustrates the need for different metrics (i.e., measures and curves) when assessing performance of a classifier.

```
## # A tibble: 10 x 3
##      prob_Yes pred_class group
##         <dbl> <fct>      <fct>
##  1    0.28   No         No
##  2    0.547  Yes        No
##  3    0.43   Yes        No
##  4    0.222  No         No
##  5    0.225  No         No
##  6    0.674  Yes        Yes
##  7    0.276  No         No
##  8    0.613  Yes        Yes
##  9    0.754  Yes        Yes
## 10    0.845  Yes        Yes
```

(a) Obtain the ROC and PR curves including the corresponding *AUCs*. Can you conclude that you have a perfect classifier?

(b) Obtain the confusion matrix and the overall threshold performance measures. Do the results you obtain support your answer to the question in part (a)?

3. For this exercise, we explore alternative calculations of *AUC* based on data in Table 3 in the article by Nahm [82, p. 29]. Recall that the relevant information was also given in Section 3.2.2.

(a) Obtain the *AUC* by (i) direct calculation, and (ii) using the **yardstick** package.

(b) You can also obtain the *AUC* by using formula (1) in Kleiman and Page [66]. Show how to do this.

4. Run the following code segment to create the data for this exercise.

```
Case <- 1:20
Class <- c("p","p","n","p","p","p","n","n","p","n",
           "p","n","p","n","n","n","p","n","p","n") %>%
         as.factor() %>% fct_rev()
Score <- c(0.9, 0.8, 0.7, 0.6, 0.55, 0.54, 0.53, 0.52,
           0.51, 0.505, 0.4, 0.39, 0.38, 0.37, 0.36,
           0.35, 0.34, 0.33, 0.3, 0.1)
Threshold <- Score
Fawcett_tb <- tibble(Case, Class, Score, Threshold)
```

Data on the first three variable are from Fawcett [35, p. 864].

(a) Obtain a graphical display showing how the number of false positives and number of true positives *separately* varies with threshold.

(b) Obtain the ROC curve using the `plot()` function and determine the AUC by direct calculation (i.e., by exploiting the geometry of the curve you obtained). Provide two checks for the value of AUC that you obtained.

(c) Obtain the PR curve and the corresponding AUC by using the **yardstick** package.

4

Comparative Analysis of Classifiers

When solving a classification problem, you would typically use the same dataset to train several different classifiers and then perform a comparative analysis of them with the same validation (e.g., test) dataset. This, of course, is an essential part of the process of classifier construction for the problems that you encounter in practice. Such analysis is also relevant when you attempt to evaluate a new classification algorithm.

One important issue you face when attempting to compare classifiers is the choice of performance metric to use in the comparison. Some questions you need to keep in mind when dealing with this issue for binary classification include the following.

- Do you have an imbalanced classification problem?

- Is your focus on predicting class labels or *positive* class membership probabilities?

- Is the *positive* class more important?

- Are the two misclassification errors (i.e., false positives and false negatives) equally important?

- When errors have unequal cost, which is more costly?

For imbalanced problems, you can find some guidance on choosing a performance measure in the flowchart given by Brownlee [10, p. 46]. The flowchart makes use of the answers to the last four questions listed above to guide you in your selection. Typically, these answers require you to take into account the business goals and/or research objectives underlying the classification problem you want to solve.

For example, assuming the usual definition of a *positive* case, minimizing occurrence of false negative errors is important in coronary artery disease prediction; see Akella and Akella [1] for the reason why. On the other hand, in spam filtering applications, it is important for *precision* to be very high in order to have good control over occurrence of false discoveries (which you can achieve by giving more emphasis to controlling the occurrence of false positives).

DOI: 10.1201/9781003518679-4

One issue that you can raise with Brownlee's flowchart is the fact that the available metrics to choose from is restricted to one of the following: *accuracy*, F_β-score for $\beta \in \{0.5, 1, 2\}$, *G*-mean, *Brier score*, and the *AUC*s for ROC and PR curves. This limitation is in part due to the focus on imbalanced classification. Another issue worth keeping in mind when making your selection from the highlighted list of measures is the fact that there are problems that can arise with some of them; see Provost et al. [91], Hand [50], and Hand and Christen [46], for example.

There are other measures that you can use when faced with problematic metrics. Some of these arise in connection with questions like the following:

- Which classifier provides the best predictive accuracy after you account for the possibility of correct prediction by chance alone?

- Which classifier provides greatest reduction in expected minimum misclassification loss?

- Which classifier has the best ability to distinguish between *positive* and *negative* cases?

The above questions lead to measures like Cohen's *kappa*, *H*-measure and *Discriminant Power*. We will revisit the use of the measures highlighted so far and others in this chapter.

To show how **R** can be brought to bear on the problem, we will ignore the measure selection issue and proceed to demonstrate how to use the software and various measures in a comparative analysis. Of course, in practice, you will usually base your analysis on a narrower selection of metrics (e.g., see Akella and Akella [1]) after taking into account the relevant factors and questions you have about your problem. However, it is important to keep in mind that the various measures deal with different aspects of performance, and empirical comparisons between classifiers measuring different aspects are of limited value [52]. This remark is noteworthy in light of studies such as those by Halimu et al. [45] and Chicco and Jurman [15]; their studies resulted in different conclusions about the relative merits of *AUC* and *mcc*.

To illustrate what is involved in a comparative analysis of competing classifiers, we revisit the Titanic survival classification problem. We have already developed two machine learners for this problem, namely, a decision tree (DT) classifier in Chapter 2 and a neural network (NN) classifier in Chapter 3. In this chapter, we train two additional classifiers, and compare the relative performance of the available classifiers with help from the measures and curves that we studied in the last two chapters. We will initially take a descriptive approach in Section 4.3, and leave it to Section 4.4 to discuss some examples on whether any observed difference in performance is statistically significant.

4.1 Competing Classifiers

In this section, we train two additional classifiers for the Titanic survival classification problem. One is a logit model (LM) classifier (a statistical learner) and the other is a random forest (RF) classifier (an ensemble learner). The competitors in our comparative analysis of classifier performance will include these two classifiers in addition to the machine learners from the last two chapters.

4.1.1 Logit Model Classifier

We briefly introduced the logit model classifier in Section 1.2.1 of the first chapter. Recall, for this classifier, *positive* class membership is determined by (1.4). When applied to the Titanic survival classification problem, it suffices to specify the relevant linear predictor $\eta(x)$, assuming that you use the usual default threshold. Given the features in the training dataset (i.e., the `Titanic_train` dataset from Section 2.2.2), this is given by

$$\begin{aligned}\eta(x) &= \beta_0 + \beta_1 \times Sexmale + \beta_2 \times Age + \beta_3 \times FamSize \\ &\quad + \beta_4 \times Pclass2nd + \beta_5 \times Pclass3rd.\end{aligned} \tag{4.1}$$

```
# Logit Model (LM) Classifier

# Check Levels of the Class Variable
Titanic_train %$% levels(Survived)
## [1] "No"  "Yes"

# Fit the Logit Model
LM_fit <- glm(Survived ~ ., data = Titanic_train,
  family = binomial)
LM_fit %>% tidy() %>% select(-std.error, -statistic)
## # A tibble: 6 x 3
##   term         estimate  p.value
##   <chr>           <dbl>    <dbl>
## 1 (Intercept)     4.63  8.63e-21
## 2 Sexmale        -2.74  3.49e-33
## 3 Age            -0.0529 2.80e- 8
## 4 FamSize        -0.181  2.70e- 2
## 5 Pclass2nd      -1.39  7.95e- 6
## 6 Pclass3rd      -2.62  2.05e-19
```

Note that categorical features in the training dataset are represented by suitable indicator (i.e., dummy coded) variables. Here, these variables are *Sexmale, Pclass2nd* and *Pclass3rd*.[1]

Estimates of the β_i's in the linear predictor are obtained when the glm() function is used to fit the logit model. The resulting estimates given in the preceding code segment are statistically significant as shown by the corresponding P-values. The trained LM classifier is determined by these estimates.

To analyze performance of the trained classifier, you need to obtain predicted probabilities and classes when it is applied to a test dataset. The LM_pv tibble created in the next code segment contains the required information.

```
# Logit Model (LM) Classifier (cont'd)

# Obtain Predictions from the LM Classifier
lin_pred <- predict(LM_fit, newdata = Titanic_test)
LM_pv <-
  tibble(
    prob_No = -lin_pred %>% plogis(),
    prob_Yes = lin_pred %>% plogis(),
    pred_class = ifelse(prob_Yes > 0.5, "Yes", "No") %>%
      as.factor(),
    Survived = Titanic_test %$% Survived
)
LM_pv %>% print(n = 3)
## # A tibble: 223 x 4
##    prob_No prob_Yes pred_class Survived
##      <dbl>    <dbl> <fct>      <fct>
## 1    0.346    0.654 Yes        Yes
## 2    0.929   0.0706 No         No
## 3    0.826    0.174 No         No
## # ... with 220 more rows
```

Next, we check the levels of pred_class and Survived to see whether they are in the required order.

```
# Check Factor Levels

LM_pv %>% keep(is.factor) %>% map(~ levels(.))
## $pred_class
## [1] "No"  "Yes"
##
## $Survived
## [1] "No"  "Yes"
```

---

[1] For some discussion on encoding of categorical variables, see Kuhn and Johnson [68, p. 94], for example.

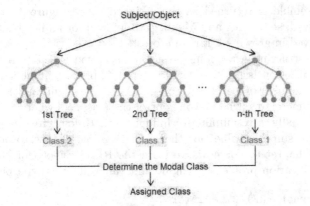

FIGURE 4.1
Illustrating a Random Forest Classifier

As can be seen, "Yes" is the second level of the two factor (i.e., categorical) variables. This is a consequence of leaving it at that level because, when fitting a logit model, the `glm()` function in the **stats** package models the logit of getting the outcome defined by the second level.

Since we'll be using **yardstick** for performance analysis, it is convenient to make "Yes" the first level for `pred_class` and `Survived` since it determines the *positive* class for our problem. This makes it the reference level and obviates the need to specify the `event_level` argument when using the functions in this package.

```
# Make Yes the First (i.e., Reference) Level

LM_pv <- LM_pv %>% mutate_if(is.factor, fct_rev)
```

4.1.2 Random Forest Classifier

Ensemble learners combine several complementary weak learners to build a more powerful classifier. Random forests (RF) classifiers, for example, are constructed from a large number of simple decision trees (these constitute the base learners); see Figure 4.1. Each decision tree in the ensemble is trained on a bootstrap sample from the training data with split variables chosen from a random subset of the original features; see Boehmke and Greenwell [6, p. 204] for an outline of the basic algorithm.[2]

When new data on a subject or object is passed through the ensemble, each base learner makes its own classification, and the modal classification

---

[2]Split-variable randomization introduces more randomness into the tree-growing process, and hence helps to reduce "tree correlation".

from the ensemble is returned as the assigned class. Figure 4.1 illustrates what is involved; see Zumel and Mount [119, p. 362] for an illustration of the process involved in growing a random forest.

When constructing a random forest classifier, you typically have several decisions to make on issues like decision tree complexity, number of trees in the ensemble, number of features to consider for a given split, and so on. There are hyperparameters that allow you control these aspects when training such a classifier. Usually, a resampling technique (e.g., 10-fold cross-validation) is used to obtain suitable values for them. An example of how to do this will be given in Chapter 6. For now, we train the RF classifier (for the Titanic survival classification problem) using default values for the hyperparameters.

```
# Random Forest (RF) Classifier

# Training the RF Classifier
library(randomForest) # help(package = "randomForest")
set.seed(290623)
RF_fit <- randomForest(Survived ~ ., data = Titanic_train,
  importance = TRUE)

# Obtain Predictions from the RF Classifier
prob_NY <- RF_fit %>% predict(newdata = Titanic_test,
  type = "prob") # a "matrix" object
RF_pv <-
  tibble(
    prob_No = prob_NY %>% .[,1],
    prob_Yes = prob_NY %>% .[,2],
    pred_class = RF_fit %>% predict(newdata = Titanic_test,
      type = "response"),
    Survived = Titanic_test %$% Survived
  )

# Make Yes the Reference (i.e., Positive) Class
RF_pv <- RF_pv %>% mutate_if(is.factor, fct_rev)
```

The variable importance plot for the RF classifier may be obtained by running the following code (figure omitted).

```
# Display the Variable Importance Plot

RF_fit %>% varImpPlot(type=2, main="Variable Importance Plot")
```

When you run the code, you'll see agreement with the DT classifier on what constitutes the most important feature for the classification problem. There is, however, disagreement on the relative importance of the remaining three features.

4.2 Collect Predictions

To facilitate comparisons of the four classifiers that we constructed for the Titanic survival classification problem, it is convenient to collect the predictions from the various classifiers into a single **R** data object. This can be done in a number of ways. Before describing two approaches, we start by importing the predictions that were obtained in the last two chapters.

```
# Import Predictions from DT and NN Classifiers

DT_pv <- read_csv("DT_pv.csv", col_types = "ddff")
NN_pv <- read_csv("NN_pv.csv", col_types = "ddff")
```

4.2.1 Using a List of Tibbles

One way to collect the predictions from the four competing classifiers is to use a list of tibbles. With such a list, you can easily extract useful information from the various tibbles with help from the `map()` function in the **purrr** package. The following code segment illustrates this.

```
# Combine Predictions Using a List (Output Omitted)

# Create the List of Predictions
pv_lst <- list(DT_pv, LM_pv, NN_pv, RF_pv)
classifiers <- c("DT", "LM", "NN", "RF") # note alpha order
names(pv_lst) <- classifiers

# Extract Some Information from the List
pv_lst %>% map(~ head(., 2))
pv_lst %>% map(~ dim(.))

# Check levels of Predicted and Actual Classes
chklevs <- . %>% keep(is.factor) %>% map(~ levels(.))
pv_lst %>% map(~ chklevs(.))
```

With pv_lst at hand, you can easily obtain array summaries, performance measures and curves for the competing classifiers. We demonstrate how to obtain the required metrics in later sections of this chapter.

4.2.2 Using a Single Tibble

Perhaps a more useful combination is to use a single tibble to contain all the predictions from the competing classifiers, particularly when you group the information in the tibble by classifiers. To obtain such a tibble, you can proceed as follows.

```
# Combine Predictions into a Single Tibble

# Create a Single Tibble Containing All Predictions
pv_tb <-
    bind_rows(DT_pv, LM_pv, NN_pv, RF_pv) %>%
    mutate(classifier = rep(classifiers, each = nrow(DT_pv))) %>%
    group_by(classifier)

# Obtain a Partial Listing of pv_tb
pv_tb %>% print(n = 3)
## # A tibble: 892 x 5
## # Groups:   classifier [4]
##    prob_No prob_Yes pred_class Survived classifier
##      <dbl>    <dbl> <fct>      <fct>    <chr>
## 1    0.423    0.577 Yes        Yes      DT
## 2    0.835    0.165 No         No       DT
## 3    1        0     No         No       DT
## # ... with 889 more rows
```

4.3   Descriptive Comparisons

We begin our comparison of the four competing classifiers for the Titanic survival classification problem with a descriptive analysis based on selected performance measures and curves that we discussed in the last two chapters. At this point, we focus on demonstrating how to use **R** to highlight differences in performance without worrying about statistical significance of the observed differences. Some discussion of the latter aspect will be taken up in Section 4.4.

4.3.1   Compare Array Summaries

One way to obtain array summaries for the competing classifiers is to produce a display of their confusion matrices like the one given in Figure 4.2. This figure was obtained using commands in the following code segment.[3]

```
# Confusion Matrices of Competing Classifiers

# Obtain List of Confusion Matrices
```

---

[3] Figure 4.2 is quite useful for publication purposes but it requires more coding effort than needed. Exercise 1 explores simpler alternative ways to get the required array summaries.

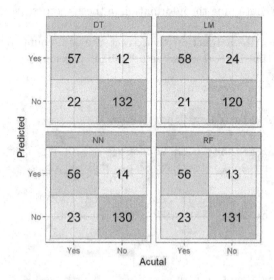

FIGURE 4.2
Tile Plots of the Confusion Matrices

```
cm_lst <- pv_lst %>%
  map(~ conf_mat(., Survived, pred_class,
      dnn = c("Predicted", "Actual")))

# Obtain Tile Plots of the Confusion Matrices
pv_lst %>%
  map(~ with(., table(pred_class,Survived)) %>% as_tibble()) %>%
  map(~ mutate_at(., .vars = c("pred_class", "Survived"),
    .funs = factor)) %>%
  map(~ mutate(., Survived = relevel(Survived, ref="Yes"))) %>%
  reduce(bind_rows) %>%
  mutate(classifier = factor(rep(classifiers, rep(4,
    each = 4)))) %>%
  ggplot(aes(x = Survived, y = pred_class, alpha = n)) +
    geom_tile(fill = "lightblue", col = "blue",
      show.legend = FALSE) +
    geom_text(aes(label = n), col = "black", alpha = 1,
      size = 5) +
    labs(x="Acutal", y="Predicted") + facet_wrap(~classifier)
```

In Chapter 1, we saw an alternative way to represent information in a *binary* confusion matrix that is simpler. This involves the use of a key count vector. Thus, as an alternative to Figure 4.2, you may use a list of key count vectors or a tibble containing the information in these vectors. We demonstrate this next.

To begin, we write a function to extract the key count vector from a confusion matrix that is represented as a `"table"` object.

```
# Compare Key Count Vectors

# Function to Obtain a Key Count Vector
kcv <- function(cm) {
  # Purpose: Obtain Key Count Vector
  # Input: Binary Confusion Matrix as "table" Object

  kc <- c(cm[1, 1], cm[2, 1], cm[1, 2], cm[2, 2])
  names(kc) <- c("TP", "FN", "FP", "TN")
  return(kc)
}

# Obtain the Key Count Vectors
cm_lst %>% # objects in this list are of type "conf_mat"
  map(~ pluck(., 1) %>% kcv()) %>%
  reduce(bind_rows) %>%
  mutate(classifier = classifiers) %>%
  select(classifier, everything())
## # A tibble: 4 x 5
##    classifier    TP     FN     FP     TN
##    <chr>      <int>  <int>  <int>  <int>
## 1 DT            57     22     12    132
## 2 LM            58     21     24    120
## 3 NN            56     23     14    130
## 4 RF            56     23     13    131
```

A quick scan of Figure 4.2 or the tibble in the above code segment shows that the DT classifier has a slight edge over the NN and RF classifiers, and the worst performing one being the LM classifier due to its relatively high number of false positives.[4] The tibble makes it easier to compare the key counts of the competing binary classifiers. The relatively similar FN values are clearly evident. We can say the same for three of the FP values, the one outlier is due to the poor performance of the LM classifier.

4.3.2   Compare Threshold Performance Measures

Threshold measures provide one perspective on relative performance of the competing classifiers, i.e., they help you compare the ability of the classifiers to correctly predict the class of a case by measuring the extent of agreement

---

[4]Note that the columns of all four confusion matrices have the same totals.

between actual and predicted classes (for a given threshold, usually the default). The relevant measures to use in the comparative analysis depend on issues like whether there is serious class imbalance, the relative cost of classification errors and so on.

A convenient way to get the measures that were discussed in Chapter 2 for the competing classifiers is to use information in the `pv_tb` from Section 4.2.2 and the `conf_mat()` function to obtain a `"conf_mat"` object for each classifier followed by application of the `summary()` function to extract the measures. The results you obtain may then be subsequently assembled into a single tibble as shown below.

```
# Compute Threshold Performance Measures

pm_tb <- pv_tb %>%
  conf_mat(Survived, pred_class) %>%
  pluck("conf_mat") %>%
  map(~ summary(.) %>% select(-.estimator)) %>%
  reduce(inner_join, by = ".metric") %>%
  set_names(c("Measure", classifiers))
```

For each classifier, there are 13 measures in `pm_tb` but note that there are some redundancies (e.g., both *ppv* and *precision* are included). A useful approach to use when you proceed to examine the measures is to look at relevant subgroups of measures. For example, you can start by looking at measures that tell you something about overall performance like those given in the next code segment and displayed in Figure 4.3. Your choice on which measure to focus on in this group depends on its interpretability, the problem, and what aspect of overall performance matters to you.

```
# Compare Overall Measures

pm_tb %>% slice(1,2,7) %T>% print() %>%
  gather(key = classifier, value = estimate, -Measure) %>%
  ggplot(aes(x = estimate, y = classifier, fill = classifier)) +
    geom_bar(stat="identity", show.legend = FALSE) +
    facet_wrap(~ Measure)
## # A tibble: 3 x 5
##   Measure       DT    LM    NN    RF
##   <chr>      <dbl> <dbl> <dbl> <dbl>
## 1 accuracy   0.848 0.798 0.834 0.839
## 2 kap        0.657 0.563 0.628 0.637
## 3 mcc        0.660 0.563 0.630 0.640
```

Usually, *accuracy* is the first measure that comes to mind. But, is it appropriate for the problem you face? This largely depends on the extent of class imbalance in your classification problem. When imbalance is not severe,

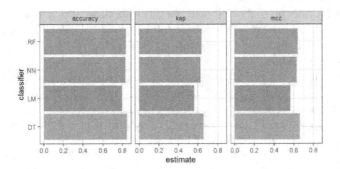

FIGURE 4.3
Bar Plots of Overall Performance Measures

accuracy does provide useful information; see Provost et al. [91] for issues involved in use of this measure. For the problem at hand, there is some degree of class imbalance but it is not severe (if it was, *accuracy* would not be a suitable measure to consider). The *accuracy* values of all the competing classifiers are significantly greater than 0.646 (this is the estimated NIR that we saw in Section 2.5.2). When compared, their values reflect the initial impression on relative performance that was obtained when we examined the confusion matrices in Section 4.3.1. The same can be said about the other two overall measures. The agreement in relative rankings of the competing classifiers shown by the three overall measures cannot be expected to occur for other comparisons. This means that you often have to make choice between these measures.

If you have to narrow your choice to one of the three overall measures, which should you choose? Chicco and Jurman [14] have argued that *mcc* is preferred to *accuracy* (and F_1-measure). Arguments against the use of *kappa* have been provided by Delgado and Tibau [22] and others. Furthermore, research by Powers [88] suggests that *mcc* is one of the best balanced measure. Given the arguments presented by these researchers, the measure that merits serious consideration is *mcc*, unless there are compelling issues in your problem that suggests otherwise.

What other groups of measures should you examine? One approach to deal with this issue is to consider complementary pairs of class-specific metrics (we consider two such pairs in the sequel). As noted by Cichosz [16], it takes a complementary pair of indicators (i.e., metrics) to adequately measure the performance of a classifier.

Often, the choice of measures depends on the relative cost of the possible classification errors. If your goal is to minimize false negatives and false positives, then the group of measures in the next example is what you probably want to examine.[5] Here, the complementary pair is *sensitivity* and

---

[5]The code you need to produce the results exhibited is similar to what was given for overall measures. It suffices to change the arguments to the `slice()` function.

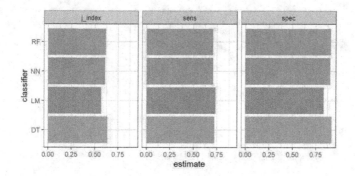

FIGURE 4.4
Bar Plots of *J*-index, Sensitivity and Specificity

specificity, and *J*-index is the composite measure that combines this pair. These measures for the competing classifiers are given below and displayed in Figure 4.4.

```
# Compare J-Index, Sensitivity and Specificity

## # A tibble: 3 x 5
##    Measure     DT     LM     NN     RF
##    <chr>    <dbl>  <dbl>  <dbl>  <dbl>
## 1 j_index  0.638  0.568  0.612  0.619
## 2 sens     0.722  0.734  0.709  0.709
## 3 spec     0.917  0.833  0.903  0.910
```

The "best performing" DT classifier that we have identified does well with *specificity* but not quite as well with *sensitivity*. On balance, as shown by the *J*-index values, this classifier does better than the others.

On the other hand, if you want to minimize false discovery errors and yet keep occurrence of false negatives as low as possible, then you may wish to consider *precision* and *recall*, and possibly F_1-measure as shown below and in Figure 4.5.

```
# Compare F1-Measure, Precision and Recall

## # A tibble: 3 x 5
##    Measure      DT     LM     NN     RF
##    <chr>     <dbl>  <dbl>  <dbl>  <dbl>
## 1 f_meas    0.770  0.720  0.752  0.757
## 2 precision 0.826  0.707  0.8    0.812
## 3 recall    0.722  0.734  0.709  0.709
```

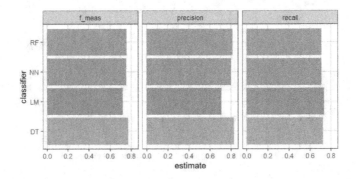

FIGURE 4.5
Bar Plots of F_1-measure, Precision and Recall

The DT classifier has the highest *precision* and the LM classifier the lowest (the difference is *descriptively* significant).[6] On the other hand, *sensitivity* or *recall* is highest for the LM classifier. Corresponding values of these measures for the NN and RF classifiers are quite comparable.

4.3.3 Compare Performance Curves and AUCs

Threshold performance measures allow you to assess only certain aspects of classifier performance. They do not, for example, capture other aspects like separability of the distributions of *positive* class membership probabilities for the two classes in a binary classification problem. Separability is important because a classifier cannot be expected to perform well if it does not provide good separation between the two distributions. For classifiers with prediction function of the form (1.2), good separability will ensure that *positive* class membership probabilities of cases in the *positive* class will more likely be ranked higher than corresponding probabilities of those in the *negative* class. In light of the material covered in the last chapter, you can compare this aspect of classifier performance by looking at the ROC curves and *AUC*s of the competing classifiers.

There is, however, one problem with what we are attempting to do and it has to do with the incoherence of *AUC* (this issue will re-surface later when we compare confidence intervals for it). Nonetheless, we continue with the comparisons for the purpose of illustrating use of **R** for the problem at hand. In any case, we also report values obtained for the *H*-measure.

The next code segment shows how to get the required ROC curves and measures from pv_lst and pv_tb that were obtained in Section 4.2.2.

---

[6]Descriptive significance just depends on visual judgement on the difference in performance. For statistical signficance, you need to account for variability in the estimated measure and evaluate the probability of getting a difference as large as what is observed.

```
# Compare ROCs, AUCs and H-Measures

# Plot the ROC Curves
rocpts_lst <- list()
for (i in 1:4) {
  rocpts_lst[[i]] <- pv_lst[[i]] %>%
    roc_curve(Survived,prob_Yes) %>%
    arrange(specificity,sensitivity) %>%
    mutate(classifier=classifiers[i] %>% as.factor())
}
rocpts_lst %>% bind_rows() %>%
  ggplot(aes(x=1-specificity,y=sensitivity,group=classifier)) +
  geom_line(size=1,aes(color=classifier,linetype=classifier)) +
  geom_segment(aes(x=0,y=0,xend=1,yend=1),linetype=2) +
  theme_bw()

# Obtain the AUCs and H-Measures
pv_tb %>% roc_auc(Survived,prob_Yes) %>%
  select(classifier,.estimate) %>%
  rename(AUC=.estimate) %>%
  inner_join(summarize(pv_tb,HM=HM_fn(Survived, prob_Yes)),
    by="classifier")
## # A tibble: 4 x 3
##    classifier    AUC    HM
##    <chr>       <dbl> <dbl>
## 1 DT          0.817 0.499
## 2 LM          0.818 0.464
## 3 NN          0.829 0.483
## 4 RF          0.856 0.541
```

Also computed are values of H-measure, the coherent alternative to AUC proposed by Hand [50]. Recall that this performance measure tells you something about the reduction in expected minimum misclassification loss that can be obtained from a trained classifier. The required values are given by HM in the above code segment. They were obtained using the HM_fn() wrapper function given in Section 3.2.2.

As can be seen from the ROC curves in Figure 4.6, the four classifiers performed significantly better than the random classifier. Another way to see this is to examine the corresponding AUCs. The areas obtained for all the competing classifiers exceed the lower threshold of 0.8 for good performance suggested by Nahm [82]. Note that the ROC curves of the competing classifiers cross one another, i.e., none of them show clear dominance in the ROC space although the curve for the RF classifier is, for the most part, above the curves for the other three classifiers. Not surprisingly, this somewhat dominant feature is also reflected in the value of its AUC. The superior performance of

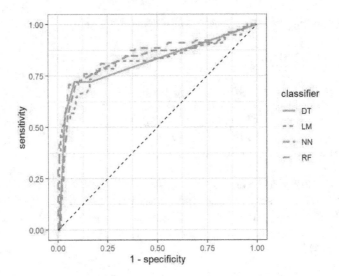

FIGURE 4.6
ROC Curves of the Competing Classifiers

the RF classifier still holds when PR curves and the corresponding AUCs are compared; see Exercise 2.

More important to note is the fact that the RF classifier provides the greatest reduction in expected minimum misclassification loss as can be seen when you compare the values of H-measure (as given by HM). Comparisons based on this alternative measure is preferred given the problems with the incoherence of AUC, and when ROC curves cross one another.

Finally, note that Figure 4.7 illustrates the separability of the various density plots for the *positive* class membership probabilities.

4.3.4 Compare Other Performance Measures

The performance measures and curves from the last two chapters are the standard ones that you would typically use to evaluate performance of one or more classifiers. There are other potentially useful metrics that you can use in the analysis. In this section, we discuss the use of two other measures for the comparative analysis of the competing classifiers.

The first alternative measure is *discriminant power* (this is also a threshold performance measure) defined by

$$DP = \frac{\sqrt{3}}{\pi} \ln DOR,$$

where

$$DOR = \frac{tpr/fnr}{fpr/tnr} = \frac{\text{odds of a \textit{positive} classification for a \textit{positive} case}}{\text{odds of a \textit{positive} classification for a \textit{negative} case}}$$

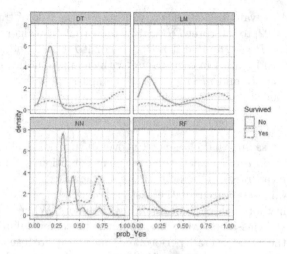

FIGURE 4.7
Distributions of Positive Class Membership Probabilities

is the *diagnostic odds ratio*. Note that DOR may be re-expressed as [102]

$$\frac{tpr/fpr}{fnr/tnr} = \frac{positive\ likelihood\ ratio}{negative\ likelihood\ ratio} = \frac{plr}{nlr}.$$

The first expression for DOR shows that, like AUC, it is also influenced by the separability of *positive* class membership distributions of the two classes. The nature of this apparent connection between the two measures is worth investigating.

```
# Compare Discriminant Power

pv_tb %>% summarize(
  tpr = sens_vec(Survived, pred_class),
  tnr = spec_vec(Survived, pred_class)
) %>%
  mutate(DOR = (tpr / (1 - tpr)) / ((1 - tnr) / tnr)) %>%
  mutate(DP = (sqrt(3) / pi) * log(DOR))
## # A tibble: 4 x 5
##   classifier   tpr   tnr   DOR    DP
##   <chr>      <dbl> <dbl> <dbl> <dbl>
## 1 DT         0.722 0.917  28.5  1.85
## 2 LM         0.734 0.833  13.8  1.45
## 3 NN         0.709 0.903  22.6  1.72
## 4 RF         0.709 0.910  24.5  1.76
```

When used to evaluate a classifier, you can use the rule of thumb given in Sokolova et al. [102] to interpret DP which states that "the algorithm is a poor discriminant if $DP < 1$, limited if $DP < 2$, fair if $DP < 3$ and good in other cases". Also, from Figure 1 in their paper (with change in notation), they gave criteria involving plr and nlr to facilitate comparison of two classifiers. For example, one rule states that classifier 1 is superior overall to classifier 2 if

$$plr_1 > plr_2 \text{ and } nlr_1 < nlr_2.$$

This amounts to saying that classifier 1 is overall superior if $DOR_1 > DOR_2$. In terms of *discriminant power*, the equivalent condition is $DP_1 > DP_2$. We discuss these measures for the competing classifiers next.

The DP values reported in the last code segment suggest limited discriminant ability for each of the competing classifiers. When this measure is used to rank the classifiers, you wind up with the same ranking as that obtained when the usual overall performance measures are used as shown in Section 4.3.2 (note that agreement also holds with J-index and F_1-measure).

The final measure we consider in this section for the comparative analysis is the *Brier score* defined by

$$BS = \frac{1}{n} \sum_{i=1}^{n} (\hat{y}_i - y_i)^2,$$

where $\hat{y}_i$ is the predicted *positive* class membership probability for the i-th case and y_i is equal to 1 or 0, according to whether the case is *positive* or not. This measure belongs to the probabilistic metric category as defined by the taxonomy that is given in Ferri et al. [37]. It is a "negative oriented" measure, which means that smaller values of the score indicate better predictions [98]. In the next code segment, we compare BS for the competing classifiers.[7]

```
# Compare Brier Scores

pv_tb %>%
  mutate(yhat = prob_Yes) %>%
  mutate(y = ifelse(Survived == "Yes", 1, 0)) %>%
  summarize(BS = sum((y-yhat)^2)/length(y))
# A tibble: 4 x 2
  classifier     BS
  <chr>        <dbl>
1 DT           0.128
2 LM           0.147
3 NN           0.166
4 RF           0.129
```

---

[7]Comparisons can also be based on the "positive oriented" Brier skill score (BSS) but this requires you to take into account the BS for a reference prediction; see Roulston [98], for example. You have to consider this adjustment if you are not willing to consider "negative oriented" measures.

The DT classifier has the smallest BS with the RF classifier a close second. The ranking of classifiers that results from BS differs from that noted for DP and several other measures.

4.4 Assessing Statistical Significance

In the last section, we obtained several (threshold, ranking, and probabilistic) performance measures for the four competing classifiers under consideration for the Titanic survival classification problem. Before taking up the question of whether observed differences between classifiers in the values of a measure are statistically significant, we summarize the results that were obtained by ranking the competing classifiers (from "best" to "worst") according to each of nine selected measures. Admittedly, the number of measures is more than what you need in practice to perform the evaluation, e.g., Akella and Akella [1] considered only four measures in their comparative analysis of six classifiers for detection of coronary artery disease.

```
# Ranking of Classifiers by Nine Performance Measures

## # A tibble: 9 x 5
##    Measure     DT     LM     NN     RF
##    <chr>    <dbl>  <dbl>  <dbl>  <dbl>
## 1 accuracy     1      4      3      2
## 2 kap          1      4      3      2
## 3 mcc          1      4      3      2
## 4 j_indec      1      4      3      2
## 5 f_meas       1      4      3      2
## 6 AUC          4      3      2      1
## 7 HM           2      4      3      1
## 8 DP           1      4      3      2
## 9 BS           1      3      4      2
```

As shown in the above results, all the threshold measures gave the same rankings to the four classifiers with the DT classifier ranked as the "best" and the LM classifier ranked as the "worst". On the other hand, the RF classifier was ranked the "best" by AUC and HM (i.e., H-measure), but there is disagreement with the ranking of the other classifiers by these two measures. In particular, AUC ranked the DT classifier as worst of the four, unlike the rankings by threshold measures (this contrarian finding is possibly a consequence of the incoherence of AUC). The ranking by BS partially agrees with that given by the threshold measures.

Of course, the usefulness of the rankings that we noted above is predicated on the assumption that differences in the values of a particular measure among the classifiers are statistically significant, but are they? In this section, we examine this issue for two measures, namely, *accuracy* and AUC by comparing confidence intervals for the corresponding probability metric. Despite issues with the measures involved here, we proceed with the comparisons for illustrative purposes. You can perform similar analysis if you can obtain confidence intervals for other suitable measures (e.g., H-measure).

4.4.1   Statistical Significance of Accuracy Differences

One approach you can take to assess the statistical significance of differences in *accuracy* values that were obtained in Section 4.3.2 for the competing classifiers is to construct confidence intervals for the probability metric in Table 2.3 that defines *Accuracy*.

In the following code segment, the required intervals are obtained using the acc_ci() function given in Appendix A.3. The resulting intervals are plotted in Figure 4.8.

```
# Compare 95% Confidence Intervals for Accuracy

# Compute and Plot the Confidence Intervals
cm_lst %>% map(~ pluck(., 1) %>% acc_ci()) %>%
  reduce(bind_rows) %>%
  mutate(classifier = classifiers) %>%
  select(classifier, everything()) %T>%
  print %>%
  ggplot(aes(estimate, fct_reorder(classifier, estimate))) +
    geom_errorbar(aes(xmin = lower_limit, xmax = upper_limit),
      width = 0.2, alpha = 0.7, col = "blue") +
    geom_point(col = "red") + labs(x = "Estimate",
      y = "Classifier") + theme_bw()
## # A tibble: 4 x 4
##    classifier estimate lower_limit upper_limit
##    <chr>         <dbl>       <dbl>       <dbl>
## 1 DT            0.848       0.794       0.892
## 2 LM            0.798       0.739       0.849
## 3 NN            0.834       0.779       0.880
## 4 RF            0.839       0.784       0.884
```

The confidence intervals displayed in Figure 4.8 suggests that the difference in *accuracy* estimates obtained for the competing classifiers are not statistically significant. It suffices to note that the *accuracy* estimates for the three other classifiers are contained in the interval for the DT classifier. This also holds when the focus is on any one of the other intervals.

The thinking involved in the above conclusion is based on the confidence interval approach to hypothesis testing. Thus, if θ denotes *Accuracy* for one of the classifiers, then $\theta - \theta_0$ is not significantly different from 0 if the specified value θ_0 is contained in the confidence interval (usually, at 95% level of confidence) for θ. Here, it suffices to let θ_0 represent one of the other estimated *Accuracy* value.

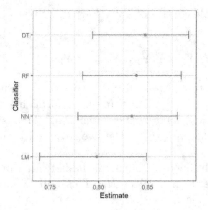

FIGURE 4.8
95% Confidence Intervals for Accuracy

As an auxiliary analysis and for purpose of illustration, consider testing the hypothesis that the DT and RF classifiers (the top two according to the rankings by most of the measures) have the same error rates. The null hypothesis for this problem may be expressed as

$$H_0 : P(\widehat{Y}_1 \neq Y, \widehat{Y}_2 = Y) = P(\widehat{Y}_1 = Y, \widehat{Y}_2 \neq Y), \qquad (4.2)$$

where $\widehat{Y}_1$ and $\widehat{Y}_2$ are the predicted target variables by the DT and RF classifiers, respectively, of the actual target Y for a case being classified. The specified null hypothesis is an alternative expression of that in Dietterich [24].

```
# McNemar's Test Comparing DT & RF Classifiers

# Actual and Predicted Classes
A <- DT_pv$Survived
P1 <- DT_pv$pred_class
P2 <- RF_pv$pred_class

# Compute the Test Statistic; see Dietterich (1998)
(n01 <- ((P1 != A) & (P2 == A)) %>% sum())
## [1] 3
(n10 <- ((P1 == A) & (P2 != A)) %>% sum())
## [1] 5
(test_stat <- (abs(n01-n10) - 1)^2 / (n01+n10))
## [1] 0.125

# Compute the P-value
pchisq(test_stat, df = 1, lower.tail = FALSE)
## [1] 0.724
```

Given its acceptable control of Type I error, we test (4.2) versus the alternative hypothesis

$$H_1 : P(\widehat{Y}_1 \neq Y, \widehat{Y}_2 = Y) \neq P(\widehat{Y}_1 = Y, \widehat{Y}_2 \neq Y)$$

by using McNemar's test. Application of this test to the problem of determining whether one learning algorithm outperforms another on a particular classification task is described in Dieterich [24].[8] This is one of the two top approximate statistical tests out of the five reviewed by him for the problem under consideration.

The P-value we obtained for our hypothesis testing problem is 0.724, and this suggests that we do not reject H_0. This null hypothesis basically says that the DT and RF clasifiers that we trained using the same dataset have the same performance (as measured by error rates). Of course, we already concluded this in light of what was observed in Figure 4.8.

4.4.2   Statistical Significance of AUC Differences

For area under the ROC curve, the relevant probability metric to consider is given by the right-hand side of (3.2). Confidence intervals for this metric may be obtained for the competing classifiers by using the approach discussed in Engelmann et al. [33], for example. However, in the following example, we use the `ci.auc()` function in the **pROC** package to obtain the required confidence intervals. By default, this function uses 2000 bootstraps to calculate an interval.

```
# Compare 95% Confidence Intervals for AUC

library(pROC)
pv_lst %>% map(~ ci.auc(.$Survived, .$prob_Yes))
## $DT
## 95% CI: 0.757-0.877 (DeLong)
##
## $LM
## 95% CI: 0.751-0.885 (DeLong)
##
## $NN
## 95% CI: 0.765-0.893 (DeLong)
##
## $RF
## 95% CI: 0.797-0.916 (DeLong)
```

---

[8]Note that this application of McNemar's test differs from that discussed in Section 2.5.2.

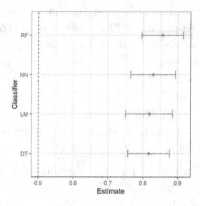

FIGURE 4.9
95% Confidence Intervals for AUC

The confidence intervals obtained in the above example are plotted in Figure 4.9. The plot shows that the AUCs of all classifiers are significantly (at 2.5% level of significance) greater than 0.5, the AUC of a random classifier. The overlapping intervals suggest that the AUCs of the classifiers do not differ significantly.

The results from the comparative analysis suggest that performance of the four classifiers that we trained for the Titanic survival problem did not differ significantly, regardless of whether assessments were based on threshold or ranking measures. This should not be taken as a conclusive finding since we did not strive to obtain the best-trained classifier in each case. We were not so motivated because our objective was to have a set of competing classifiers to illustrate techniques for a comparative analysis (recall, a similar thinking was also considered when classifiers where trained for binary CPA in the previous two chapters). Thus, we did not pay adequate attention to important aspects like feature engineering and hyperparameter tuning (the latter will be taken up in Chapter 6). It is perhaps possible that one of the four classifiers under consideration will dominate when a more comprehensive approach to training is undertaken for these classifiers. We leave this to the reader to explore. It will require more knowhow than what we have been able to cover in this book.

4.5 Exercises

1. You saw one way to obtain a display of the confusion matrices for competing classifiers in Section 4.3.1; see Figure 4.2. There are simple quicker alternatives that you can use to obtain the required array summaries. These alternatives are explored in this exercise.

(a) Show how to use `pv_lst` from Section 4.2.1 and the `con_mat()` function in Appendix A.3 to obtain a list of confusion matrices for the competing classifiers (for display on the console).

(b) You can produce heat maps of the confusion matrices in the list obtained in part (a). Show how to do this (the `do.call()` function and the **gridExtra** package are useful for this purpose).

2. In this exercise, we will tie up some "loose ends" that involve omitted code for some results presented in this chapter, and other analysis you can perform.

(a) Show how to use `pv_tb` from Section 4.2.2 and the **ggplot2** package to obtain Figure 4.7

(b) Show how to obtain Figure 4.9. You may assume the limits of the confidence intervals given in Section 4.4.2.

(c) Compare the PR curves for the competing classifiers considered in this chapter and obtain the corresponding AUCs. Comment on the results you obtain.

3. To illustrate discussion in their paper, Sokolova et al. [102] considered the use of traditional and discriminant measures to compare the performance of two classifiers (support vector machine versus naive Bayes).[9] The confusion matrices involved in the comparison was given in Table 2 of their paper.

(a) Verify the values of the four threshold measures given in Table 3 of Sokolova et al. [102].

(b) Compare the *discriminant power* of the two classifiers

(c) What do the results in parts (a) and (b) tell you about the relative merits of the two classifiers?

4. Customer churn in a business offering cell phone service occurs when customers switch from one company to another at the end of their contract. One way to manage the problem is to make special offers to retain customers prior to expiration of their contracts (since recruiting new customers to replace those that are lost is more costly). Identifying customers to make the offers to gives rise to a classification problem.

This exercise is based on confusion matrices reported in Provost and Fawcett [90, p. 227] for two classifiers in connection with such a classification problem; assume those planning to churn belong to the *positive* class. The corresponding key count vectors for the naive Bayes (NB) and k-nearest neighbors (k-NN) classifier are given in the following tibble.

---

[9]See Chapter 6 of Rhys [95] for some information on these classifiers.

```
## # A tibble: 2 x 5
##   classifier    tp    fn    fp    tn
##   <chr>      <dbl> <dbl> <dbl> <dbl>
## 1 NB           127   200   848  3518
## 2 kNN            3   324    15  4351
```

Note that the given key counts are actually from one fold in the ten-fold cross-validation analysis done by the authors based on data from the KDD Cup 2009 churn problem.

(a) Obtain heat maps of the confusion matrices.

(b) Compare the overall threshold measures. Is *accuracy* a suitable measure to use in the comparison?

(c) Compare the F_2-measure for the classifiers. Is this a suitable measure to use?

(d) Compare the *discriminant power* of the classifiers. Comment on the *dp* values you obtain.

5. The following tibble contains key count vectors from five confusion matrices given in Baumer et al. [5].

```
## # A tibble: 5 x 5
##   classifier    tp    fn    fp    tn
##   <chr>      <dbl> <dbl> <dbl> <dbl>
## 1 DT          3256  3061   990 18742
## 2 kNN         3320  2997   857 18875
## 3 NB          2712  3605  1145 18587
## 4 NN          4189  2128  1861 17871
## 5 RF          4112  1261  2205 18471
```

The labels used for the classifiers refer to decision tree (DT), *k*-nearest neighbors (kNN), naive Bayes (NB), neural network (NN) and random forest (RF). The classifiers were for the problem of classifying whether an individual is a high-income earner (such individuals belong to the *positive* class).

An important point to note about the given key counts is the fact that they were derived from *training* data rather than test data. This fact should be of concern if one is interested in generalizability of the classifiers, but we will ignore it since the purpose of this exercise is to demonstrate techniques for the following comparisons in a comparative analysis of the five classifiers.

(a) Compare the overall performance measures.

(b) Compare *sensitivity, specificity*, and *J*-index.

(c) Compare *precision, recall*, and F_1-measure.

6. The hypothyroid classification problem given in Tartar [107] is concerned with determining whether a patient has a thyroid problem or not. The author used part of the data in the `Hypothyroid.csv` file mentioned in his book to train five classifiers, and used *accuracy* to compare their performance. We re-trained the five classifiers using a different training sample and obtained the following key counts from the test dataset.

```
##       TP FN FP   TN
## DT   32  2  5 561
## LM   24  2 13 561
## NB   26  4 11 559
## NN   29  6  8 557
## SVM 27  2 10 561
```

When applicable, use suitable graphical displays to compare the performance measures for this exercise.

(a) Compare the overall error rate and the class-specific error rates (i.e., *fnr* and *fpr*).

(b) Compare *mcc* for the classifiers.

(c) Comment on the results obtained for parts (a) and (b).

Obtain a variable tree display of true and false rates for the best performing classifier.

5

Multiclass CPA

Classification problems that involve more than two classes are also quite common. You saw examples of some substantive questions that give rise to such problems in the first chapter. As with binary problems, classifiers for a multiclass problem may be based on statistical, machine, or ensemble learning, and techniques to evaluate them are to some extent similar to what was covered earlier for binary classifier performance analysis.

You can expect some issues when you attempt to apply techniques learned for binary CPA to problems that involve more than two classes. This is not surprising because the increased number of classes presents conceptual, definitional and visualization problems when dealing with multiclass performance measures and surfaces.

Conceptually, overall measures like *accuracy*, *kappa*, and *mcc* that we encountered earlier remain unchanged when you consider them for multiclass CPA. However, for the last two measures, you need to consider more general definitional formulas since the definitions for binary classifiers usually involve the key counts from Table 1.1, e.g., the *mcc* formula given by (2.5) for binary classifiers is given in terms of these counts. The required generalizations are discussed in Section 5.2.2.

Class-specific measures like *sensitivity* and *specificity* are intrinsically binary performance metrics. Recall, alternative terminology for these measures are *true positive rate* and *true negative rate*, respectively, and definition of these rates relies on the *positive* versus *negative* class dichotomy. You can also encounter use of the term *sensitivity* when analyzing performance of multiclass classifiers. When used in this context, think of it as an estimate obtained by suitable averaging of *tpr* values from an OvR collection of binary confusion matrices. This was illustrated during our overview of multiclass CPA in the first chapter. Extensions to other class-specific measures will be discussed in Section 5.2.1.

Other issues also arise when you attempt ROC analysis for multiclass classifiers. When the number of classes is large, you face visualization problems because of the high-dimensional hypersurfaces that arise in the analysis. A related issue is calculation of the volume under these surfaces. We discuss some alternatives in Section 5.3 like the ROC curves from a class reference formulation [35] for multiclass ROC and the M-measure [56] for multiclass *AUC*.

DOI: 10.1201/9781003518679-5

We begin in the next section with a 3-class problem concerned with prediction of diabetic status of an individual. This problem will be used as a running example throughout this chapter. In addition to data preparation, training, and predictions, we discuss some techniques for data exploration (so far, this aspect of the analysis has been ignored). For most of the remaining sections, we focus on techniques to deal with the problems that were highlighted in the above introduction. In the last section, we consider some inferential aspects for performance parameters in the multiclass context.

5.1   Diabetes Classification Problem

To illustrate the various CPA techniques for multiclass problems in this chapter, we will use part of the Reaven and Miller [94] diabetes dataset. One source for the data is the **rrcov** package. As noted in the documentation for this package, the dataset contains five measurements on each of 145 non-obese adult patients classified into the following three groups: "normal", "chemical" diabetic and "overt" diabetic. The categorical target variable group for the multiclass problem that we will consider has three levels defined by these groups. The feature variables that we will use as predictors for the problem include the three primary variables, namely, glucose (measuring glucose intolerance), insulin (measuring insulin response to oral glucose), and sspg (measuring insulin resistance).

Data preparation and exploration will be considered in the next two sections and training of a multinomial logistic (ML) classifier will be taken up in Section 5.1.3. The trained classifier will be used in the subsequent discussion on performance measures and curves for multiclass classifiers.

5.1.1   Data Preparation

We begin by retrieving the diabetes data frame from the **rrcov** package and follow this by creating the diabetes_tb tibble to contain the relevant data we need for the classification problem under consideration.

```
# Data Preparation

# Get diabetes from the rrcov Package
data(diabetes, package = "rrcov")

# Create the diabetes_tb Tibble
diabetes_tb <- diabetes %>%
  select(glucose, insulin, sspg, group) %>%
  as_tibble()
```

A partial listing of `diabetes_tb` shows that we have data on three features and one target variable for 145 individuals. This is a rather small dataset, but it will serve our purpose in this chapter.

```
# Data Preparation (cont'd)

diabetes_tb %>% print(n = 3)
## # A tibble: 145 x 4
##    glucose insulin  sspg group
##      <int>   <int> <int> <fct>
## 1      356     124    55 normal
## 2      289     117    76 normal
## 3      319     143   105 normal
## # ... with 142 more rows
```

Next, we use 70% of data in `diabetes_tb` for training and reserve the rest for testing. As in the second chapter, data splitting is done with help from functions in the **rsample** package. Note the use of the `strata` argument in the `initial_split()` function to specify the variable involved in deciding how splitting is accomplished with stratification (this is usually done when class imbalance is serious enough).

```
# Data Preparation (cont'd)

# Create the Training Dataset
library(rsample)
set.seed(29622)
diabetes_split <- initial_split(diabetes_tb, prop = 0.7,
   strata = group)
diabetes_train <- diabetes_split %>% training()
diabetes_train %>% print(n = 3)
## # A tibble: 101 x 4
##    glucose insulin  sspg group
##      <int>   <int> <int> <fct>
## 1      478     151   122 chemical
## 2      439     208   244 chemical
## 3      429     201   194 chemical
## # ... with 98 more rows
```

5.1.2   Data Exploration

A 3-D scatterplot of the training dataset is given in Figure 5.1. This plot not only displays the relationship among the three feature variables but also shows how their values jointly vary for each of the three classes. The plot shows that, together, these variables have good potential for predicting diabetic status.

```
# 3-D Scatter Plot Display of the Training Data

library(scatterplot3d)
brg <- c("steelblue", "red", "green")
brg_vec <- brg[as.numeric(diabetes_train$group)]
pts <- c(16, 0, 17)
pts_vec <- pts[as.numeric(diabetes_train$group)]
diabetes_train %>%
  select(-group) %>%
  scatterplot3d(pch = pts_vec, color = brg_vec)
legend("top", legend = levels(diabetes_train$group), col = brg,
       pch = c(16,0,17), inset = -0.22, xpd = TRUE, horiz = TRUE)
```

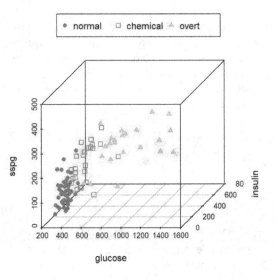

FIGURE 5.1
3-D Scatter Plot Display of the Training Data

```
# Boxplots of the Feature Variables

diabetes_train %>%
  gather(key = "Variable", value = "Value", -group) %>%
  ggplot(aes(Value, group)) +
    facet_wrap(~ Variable, scales = "free_x") +
    geom_boxplot(col = "blue", fill = "lightblue") +
    labs(x = "") + theme_bw()
```

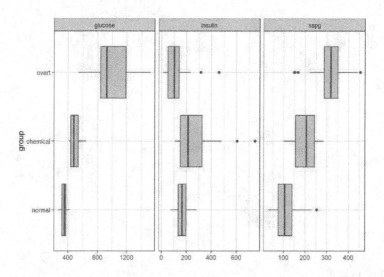

FIGURE 5.2
Boxplots of the Feature Variables

The preceding code segment yields the marginal distribution of the feature variables for each level of `group` as shown by the boxplots in Figure 5.2. The plots provide some insight on the predictive value of each feature. For example, we expect `sspg` to have good predictive value since its distribution for the three classes is quite well separated. The same can also be said about `glucose`.

As shown in Figure 5.3, the correlation between `sspg` and `glucose` is 0.79. Although quite high, the (absolute) correlation is not high enough to warrant dropping one of these two features from the training dataset.[1] Correlation between `insulin` and `glucose` is weakly negative, and that with `sspg` and is non-existent.

```
# Correlation Plot of the Feature Variables

library(corrplot)
diabetes_train %>%
  select(-group) %>%
  cor() %>%
  corrplot.mixed(lower = "ellipse", upper = "number",
    tl.pos = "lt")
```

Finally, as shown by the pie chart in Figure 5.4, prevalence of the "normal"

---

[1]Note that when there are many numeric predictors, you can use a function like `step_corr()` in the **recipes** package to identify predictors for removal from those that result in absolute pairwise correlations greater than a pre-specified threshold.

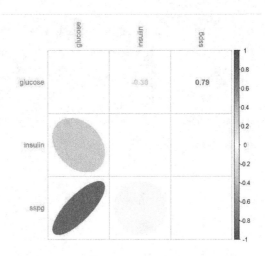

FIGURE 5.3
Correlation Plot of the Feature Variables

class is about 52.5% (there is some imbalance in the data but it is not severe).
This class determines the reference level since it is the first one for the `group`
variable.

```
# Pie Chart of the Class Variable

diabetes_train %>%
  count(group) %>%
  mutate(pct=100*n/nrow(diabetes_train)) %>%
  ggplot(aes(x="", y=pct, fill=group)) +
    scale_fill_manual(values=c("#85C1E9","#E74C3C","#ABEBC6")) +
    coord_polar("y") +
    geom_bar(width=1, size=1, col="white", stat="identity") +
    geom_text(aes(label=paste0(round(pct, 1), "%")),
              position=position_stack(vjust=0.5)) +
    labs(x=NULL, y=NULL, title="") + theme_bw() +
    theme(axis.text=element_blank(), axis.ticks=element_blank())

# Check the Reference Level
diabetes_tb %$% levels(group)
## [1] "normal"   "chemical" "overt"
```

The graphical displays in this section constitute a tiny fraction of what is
available in **R** and its packages. Though such displays are primarily used to
visualize the available features and target variable prior to training a classifier,
with some imagination, one can also employ them as part of CPA.

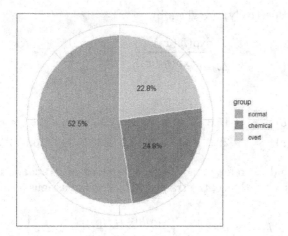

FIGURE 5.4
Pie Chart of the Target Variable

5.1.3 Training the ML Classifier

We considered a multinomial logistic (ML) classifier in Section 1.2.2 to illustrate use of a statistical learner to solve a multiclass problem. Here, we use the `multinom` function in the **nnet** package to train such a classifier for the 3-class diabetes classification problem. Prior to fitting the model, we normalize the numeric feature variables so that the range of each variable is transformed to the interval $[0, 1]$.

```
# Fit the Mutinomial Logistic (ML) Model

# Normalize Features to [0, 1] Range
library(recipes) # help(package = "recipes")
Train_data <-
  recipe(group ~ ., data = diabetes_train) %>%
  step_range(glucose, insulin, sspg) %>% prep() %>%
  juice()

# Fit the ML Model
library(nnet) # help(package = "nnet")
ML_fit <- multinom(group ~ ., data = Train_data)
ML_fit %>% summary() %>% coef()
##           (Intercept) glucose insulin  sspg
## chemical       -15.3     101  0.7867  9.01
## overt          -29.0     141 -0.0838 16.66
```

In light of the above results, the estimated probabilities of class membership (note that class "normal", "chemical" and "overt" are referred to as class

1, 2, and 3, respectively) are given by

$$\frac{\exp(\hat{\eta}_i(\boldsymbol{x}))}{\sum_{j=1}^{3} \exp(\hat{\eta}_j(\boldsymbol{x}))}, \quad i = 1, 2, 3,$$

where $\hat{\eta}_1(\boldsymbol{x}) = 0$,

$$\begin{aligned}
\hat{\eta}_2(\boldsymbol{x}) &= -15.3 + 101 \times x_1 + 0.7867 \times x_2 + 9.01 \times x_3, \\
\hat{\eta}_3(\boldsymbol{x}) &= -29.0 + 141 \times x_1 - 0.0838 \times x_2 + 16.66 \times x_3,
\end{aligned}$$

and x_1, x_2, and x_3 refer to normalized versions of `glucose`, `insulin`, and `sspg`, respectively.[2] Hence, the trained ML classifier assigns a case to the h-th class if

$$h = \underset{i}{\operatorname{argmax}} \left\{ \frac{\exp(\hat{\eta}_i(\boldsymbol{x}))}{\sum_{j=1}^{3} \exp(\hat{\eta}_j(\boldsymbol{x}))}, \; i = 1, 2, 3 \right\}, \tag{5.1}$$

or, equivalently, if

$$h = \underset{i}{\operatorname{argmax}} \left\{ \hat{\eta}_i(\boldsymbol{x}), \; i = 1, 2, 3 \right\}.$$

5.1.4   Predictions from the ML Classifier

To evaluate the trained ML classifier, we begin by preparing data for testing. Getting `diabetes_test` from `diabetes_tb` as part of data splitting is only the first step since we also need to perform some preprocessing similar to what was done for training data. As shown below, the result we get is `Test_data`. Note that we used the `bake()` function from the **recipes** package rather than the `juice()` function to extract the test data from the **recipes** object that resulted from preprocessing.

```
# Create the Test Dataset

diabetes_test  <- diabetes_split %>% testing()
Test_data <-
   recipe(group ~ ., data = diabetes_train) %>%
   step_range(glucose, insulin, sspg) %>% prep() %>%
   bake(diabetes_test)
```

For a complete performance analysis of the trained multiclass ML classifier, we need information on actual classes for each case in the test dataset in addition to corresponding hard class predictions, and for each class, predicted class membership probabilities.

---

[2]In practice, we need to examine the corresponding P-values to assess the statistical significance of the estimated coefficients, and decide what to do with those that are not significant. Here, we will ignore this issue since our purpose is to use the fitted model to illustrate multiclass CPA.

The required information is contained in the ML_pv tibble. As shown in the partial listing of this tibble given below, the columns labeled prob_normal, prob_chemical and prob_overt contain the required class membership probabilities, and the columns labeled pred_class and group contain the predicted and actual classes, respectivly.

```
# Predictions with Test Data and the ML Classifier

ML_pv <-
  bind_cols(
    ML_fit %>% predict(newdata = Test_data, type = "probs") %>%
      as_tibble() %>%
      set_names(c("prob_normal","prob_chemical","prob_overt")),
    tibble(
      pred_class = ML_fit %>% predict(newdata = Test_data,
        type = "class"),
      group = Test_data %$% group
    )
  )
ML_pv %>% print(n = 3)
## # A tibble: 44 x 5
##    prob_normal prob_chemical prob_overt pred_class group
##          <dbl>         <dbl>      <dbl> <fct>      <fct>
## 1        1.00    0.00000356   1.45e-11 normal     norm~
## 2        0.961   0.0394       1.42e- 5 normal     norm~
## 3        1.00    0.000429     7.84e- 9 normal     norm~
## # ... with 41 more rows
```

5.1.5 Array Summaries

As noted in Table 1.6 of the first chapter, the confusion matrix for a k-class problem is $k \times k$ cross-tabulation of data on predicted and actual classes. There are other ways of presenting the array summary. Other versions differ in format, labeling, and representation, e.g., see Markoulidakis et al. [77, p. 4].

A heat map of the confusion matrix for the trained ML classifier that we obtained earlier is shown in Figure 5.5. The very good performance of the classifier is clearly evident since only 3 out of the 44 cases (ignore, for the moment, the small numbers involved) in the test dataset were misclassified; the overall error rate is about 6.8%. Rather than declare that we have an acceptable classifier for the problem at hand, it is probably a good idea to subject its performance to further analysis. We'll take this up in subsequent sections.

```
# ML Confusion Matrix for the ML Classifier

library(yardstick)
ML_pv_cm <- ML_pv %>% conf_mat(group, pred_class,
  dnn = c("Predicted", "Actual"))
ML_pv_cm %>% autoplot(type = "heatmap")
```

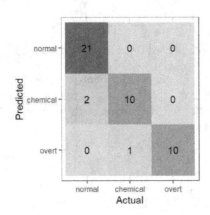

FIGURE 5.5
Confusion Matrix for the ML Classifier

You can also obtain a confusion matrix from the indicator matrices associated with the predicted and actual classes for the cases in the evaluation dataset (surprisingly, this fact is not widely known). We illustrate the approach by re-computing the confusion matrix for the ML classifier.[3]

```
# Confusion Matrix from Indicator Matrices

# Create List of Indicator Matrices
grps <- c("normal", "chemical", "overt")
XY_lst <-
  list(
    grps %>% map(function(u){ML_pv %$%
      ifelse(pred_class == u, 1, 0)}),
    grps %>% map(function(u){ML_pv %$%
      ifelse(group == u, 1, 0)})
  ) %>%
  map(~ bind_cols(.) %>% set_names(grps) %>% as.matrix())
```

---

[3]Note that in the demonstration, you can also use a function from the **ncpen** package to obtain the required indicator matrices.

```
# Confusion Matrix from Indicator Matrices (cont'd)

# Use the Indicator Matrices to obtain the Confusion Matrix
X <- XY_lst[[1]] # indicator matrix for pred_class
Y <- XY_lst[[2]] # indicator matrix for group
t(X) %*% Y # ML confusion matrix
##             normal chemical overt
## normal          21        0     0
## chemical         2       10     0
## overt            0        1    10
```

As demonstrated in the above code segment, the required array summary may be obtained from the matrix product $X'Y$ where $X = [X_{ij}]_{n\times k}$ and $Y = [Y_{ij}]_{n\times k}$ are indicator matrices that are defined by letting X_{ij} (Y_{ij}) equal to 1 or 0 according to whether or not the predicted (actual) class of the i-th case in the evaluation dataset belongs to the j-th class. The usual understanding when dealing with threshold performance measures is that they are functions of the counts in a confusion matrix. However, fundamentally, we see that these measures may also be thought of as functions of indicator matrices X and Y. As we'll see later, this insight allows us to simplify the multiclass formula for Matthews correlation coefficient.

By taking an OvR perspective, you can obtain an OvR collection of confusion matrices from the k-class confusion matrix given in Table 1.6. The i-th binary array summary in the collection is given in Table 5.1. Here, the i-th level of the target class variable for the multiclass problem plays the role of the *positive* class. The number of true positives and false negatives are given by

$$tp_i = n_{ii} \text{ and } fn_i = \sum_{r=1}^{k} n_{ri} - n_{ii},$$

respectively, and the number of false positives and true negatives are given by

$$fp_i = \sum_{r=1}^{k} n_{ir} - n_{ii} \text{ and } tn_i = \sum_{r=1}^{k}\sum_{s=1}^{k} n_{rs} - tp_i - fn_i - fp_i,$$

respectively.

Alternatively, you can consider the OvR collection of key count vectors given by

$$\{kcv_i : kcv_i = (tp_i, fn_i, fp_i, tn_i), \quad i = 1, \ldots, k\}.$$

Here, kcv_i contains the key counts from the i-th binary confusion matrix, i.e., they are those given in Table 5.1.

Next, we use the kcv_fn() function in Appendix A.3 to obtain the OvR collection of key counts from the 3-class confusion matrix displayed in Figure 5.5. The corresponding OvR collection of confusion matrices is given in Figure 5.6.

TABLE 5.1

The *i*-th OvR Binary Confusion Matrix

	Actual	
Predicted	Yes	No
Yes	tp_i	fp_i
No	fn_i	tn_i

```
# OvR Collection of Binary Array Summaries

# Obtain "table" Version of ML_pv_cm
ML_cm <- ML_pv_cm %>% pluck(1)

# Obtain OvR Collection of Key Count Vectors
classes <- c("normal", "chemical", "overt")
ML_kcv <- 1:nrow(ML_cm) %>% map(~ kcv_fn(ML_cm, .))
names(ML_kcv) <- classes

# Obtain OvR Collection of Confusion Matrices
ML_OvR_cm <- list()
for (i in 1:3){
  ML_OvR_cm[[i]] <- ML_kcv[[i]] %>%
    con_mat(c(classes[[i]],"rest"))
}
library(gridExtra)
do.call("grid.arrange", c(ML_OvR_cm %>% map(~ autoplot(.,
  type = "heatmap")), ncol = 3))
```

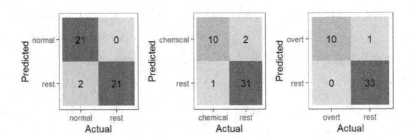

FIGURE 5.6

OvR Collection of Binary Confusion Matrices

In the next section, we use the collection of array summaries in Figure 5.6 to obtain estimates of some performance measures for multiclass CPA.

5.2 Threshold Performance Measures

To use class-specific measures like *sensitivity* and *specificity* for multiclass problems, it helps if you adopt a One-vs-Rest (OvR) perspective, for example. In our overview of multiclass CPA in Chapter 1, you saw how to exploit this perspective to obtain estimates of multiclass *sensitivity*.

To estimate multiclass versions of other class-specific measures, you can use formulas in Sokolova and Lapalme [103] or Takahashi et al. [106], for example. You can, of course, also make use of functionality in the **yardstick** package to obtain the required estimates. However, there are variations in formulas used to compute some of these estimates; for example, there are two ways to obtain macro estimates of F_1-measure. Hence, there is a need for some familiarity with formulas that are used by packaged functions.

With some exceptions, the formulas provided in the above-cited references are special cases of the formulation given in the next section for estimating class-specific measures by macro/micro averaging. We also refer to the resulting estimates as class-specific measures since they are derived from OvR collections of such estimates. Admittedly, this obscures our earlier usage of the terminology, but it suffices to take note of the number classes in the classification problem to know whether the measure is intrinsically class-specific or one obtained by averaging such measures. Another issue we'll address is the modifications to the formulas for *kappa* and *mcc* so that they are applicable to binary and multiclass problems. This will be done in Section 5.2.2.

5.2.1 Class-Specific Measures

As noted earlier, multiclass versions of binary class-specific measures may be estimated once we have an OvR collection of binary confusion matrices or the corresponding key count vectors. Assuming we have the latter, the macro averaged performance estimates for a k-class classifier may be expressed as the linear combination

$$w_1 \times pm(kcv_1) + w_2 \times pm(kcv_2) + \cdots + w_k \times pm(kcv_k),$$

where kcv_i is the i-th key count vector in the OvR collection, pm is a function that maps a key count vector to a class-specific performance measure, and w_i is the i-th weight. For example, the pm function for *sensitivity* is defined by the mapping $kcv \mapsto tp/(tp + fn)$, hence the macro and macro-weighted average estimates of *sensitivity* may be expressed as

$$w_1 \times \frac{tp_1}{tp_1 + fn_1} + w_2 \times \frac{tp_2}{tp_2 + fn_2} + \cdots + w_k \times \frac{tp_k}{tp_k + fn_k}.$$

For macro averaging, the weights are equal, but for macro-weighted averaging, they are determined by the prevalence of (i.e., fractions in) each class in the evaluation dataset.

Multiclass measures by micro averaging are given by $pm(kc\_agg)$ where kcv_agg is obtained by aggregating the key count vectors, i.e.,

$$kcv\_agg = \sum_{i=1}^{k} kcv_i = \left(\sum_{i=1}^{k} tp_i, \sum_{i=1}^{k} fn_i, \sum_{i=1}^{k} fp_i, \sum_{i=1}^{k} tn_i \right).$$

For example, the estimate of *sensitivity* by micro averaging is given by

$$\frac{\sum_{i=1}^{k} tp_i}{\sum_{i=1}^{k} tp_i + \sum_{i=1}^{k} fn_i}.$$

To obtain estimates of multiclass *sensitivity* by macro, macro-weighted and micro averaging, we begin by creating the `OvR_kcv` tibble to contain the key count vectors from the OvR collection of binary confusion matrices derived from the 3-class confusion matrix `ML_cm` for the ML classifier.

```
# Sensitivity Estimates for the ML Classifier

# OvR Collection of Key Count Vectors
OvR_kcv <-
  tibble(
    group = colnames(ML_cm),
    tp = diag(ML_cm),
    fn = colSums(ML_cm) - tp,
    fp = rowSums(ML_cm) - tp,
    tn = sum(ML_cm) - tp - fn - fp
  )
```

Next, we adapt the approach taken in Kuhn and Silge [69, p. 119] to obtain the required estimates.

```
# Sensitivity Estimates for the ML Classifier (cont'd)

# Compute macro, macro-weighted and micro Sensitivity
OvR_kcv %>%
  mutate(
    total = tp + fn,
    weight = ML_cm %>% margin.table(2) %>% prop.table(),
    sens = tp / total
  ) %T>%
  print() %>%
  summarize(
    macro = mean(sens),
    macro_wt = weighted.mean(sens, weight),
    micro = sum(tp) / sum(total)
  )
```

The results from commands in the preceding code segment are given below.

```
# Sensitivity Estimates for the ML Classifier (cont'd)

## # A tibble: 3 x 8
##   group       tp    fn    fp    tn total weight   sens
##   <chr>    <int> <dbl> <dbl> <dbl> <dbl> <table>  <dbl>
## 1 normal      21     2     0    21    23 0.523    0.913
## 2 chemical    10     1     2    31    11 0.250    0.909
## 3 overt       10     0     1    33    10 0.227    1
## # A tibble: 1 x 3
##    macro macro_wt micro
##   <dbl>    <dbl> <dbl>
## 1 0.941    0.932 0.932
```

The above *sensitivity* estimates for the ML classifier may also be obtained by using functions in the **yardstick** package. The functions in this package use the standard approach to obtain macro/micro estimates (the alternative calculation of macro F_1-measure given later illustrates use of a non-standard approach). To demonstrate the use of functions in this package, consider how to obtain macro/micro average estimates of *specificity* (the approach used to obtain multiclass *sensitivity* estimates in the preceding code segment may also be used to obtain estimates of multiclass *specificity*; see Exercise 2).

```
# Specificity Estimates for the ML Classifier

ML_pv %>% spec(group, pred_class, estimator = "macro")
## # A tibble: 1 x 3
##    .metric .estimator .estimate
##   <chr>   <chr>          <dbl>
## 1 spec    macro          0.970

ML_pv %>% spec(group, pred_class, estimator = "macro_weighted")
## # A tibble: 1 x 3
##    .metric .estimator      .estimate
##   <chr>   <chr>               <dbl>
## 1 spec    macro_weighted      0.978

ML_pv %>% spec(group, pred_class, estimator = "micro")
## # A tibble: 1 x 3
##    .metric .estimator .estimate
##   <chr>   <chr>          <dbl>
## 1 spec    micro          0.966
```

Estimates of *sensitivity* and *specificity* are commonly used in medical applications, for example, but for text classification in information retrieval, measures like *precision*, *recall*, and F_1-measure are popular [106].[4]

```
# Macro Averaged Precision, Recall & F1-Measure

ML_pv_cm %>% summary() %>% slice(11:13)
## # A tibble: 3 x 3
##    .metric   .estimator .estimate
##    <chr>     <chr>          <dbl>
## 1 precision macro          0.914
## 2 recall    macro          0.941
## 3 f_meas    macro          0.925
```

The estimate of F_1-measure for the ML classifier is given by f_meas in the above tibble. It is obtained by using the standard macro-averaging approach that we discussed earlier. There are researchers like Sokolova and Lapalme [103] who take a different view on how macro/micro estimates of the F_1-measure are defined. In their approach, the required estimate is obtained by taking the harmonic mean of the corresponding estimates for *precision* and *recall*. As noted by Takahashi et al. [106] and others, it is a less frequently used approach.

```
# Alternative Approach to Macro Average F1-Measure

pre <- 0.914; rec <- 0.941
2*(pre*rec)/(pre+rec)
## [1] 0.927
```

The estimates of multiclass versions of class-specific measures that we obtained so far provide further support for the impression we had earlier on performance by the ML classifier. Such support is necessary but not sufficient. We need further evidence from other measures like those in the next section and the performance curves that we'll examine later.

5.2.2 Overall Measures

There are measures for multiclass classifiers that do not require macro/micro averaging. These are measures that may be defined in a way that make them *directly* applicable to binary and multiclass classifiers. For example, *accuracy* as the proportion of cases that contribute to the counts on the diagonal of a confusion matrix is applicable no matter the size of this square array. For the array summary in Figure 5.5, we see that this proportion is 0.932 for the ML classifier.

---

[4]In Takahashi et al. [106] and some other references, F_1-measure is referred to as F_1-score.

Other generally applicable measures include Cohen's *kappa* and Matthews correlation coefficient *mcc*. These measures depend on certain totals that one can obtain from a confusion matrix. These include the total count, marginal column and row totals, and the sum of diagonal entries; see the formulas for $\hat{A}$ and $\hat{E}$ given below and that given by (5.3) for *mcc*. For example, consider *kappa*. As a descriptive measure, it is defined by the ratio $(\hat{A} - \hat{E})/(1 - \hat{E})$ where

$$\hat{A} = \frac{1}{n} \sum_{i=1}^{k} n_{ii} \text{ and } \hat{E} = \frac{1}{n^2} \sum_{i=1}^{k} n_{i.}n_{.i}.$$

Here, n is the total number of cases in the evaluation dataset, n_{ii} is the i-th count on the diagonal of a $k \times k$ confusion matrix, and $n_{i.}$ ($n_{.i}$) is the total count in the i-th row (column) of this array summary. As we'll see in Section 5.4.2, we can also view $\hat{A}$ and $\hat{E}$ as the maximum likelihood estimators of parameters that represent classification accuracy under different assumptions about the dependence relationship between the predicted and actual target (i.e., response) variables.

```
# Compute Cohen's kappa for the ML Classifier

# Compute Usual Accuracy Measure
CM <- ML_pv_cm %>% pluck(1) # as "table" object
(A <-  CM %>% prop.table() %>% diag() %>% sum())
## [1] 0.932

# Compute Accuracy Under Chance Agreement
ni. <- CM %>% rowSums(); n.i <- CM %>% colSums()
n <- CM %>% sum()
(E <-  sum(ni. * n.i) / n^2)
## [1] 0.374

# Compute kappa Using Definition
(A - E) / (1 - E)
## [1] 0.891

# Compute kappa Using yardstick Package
ML_pv %$% kap_vec(group, pred_class)
## [1] 0.891
```

The situation with the Matthews correlation coefficent (*mcc*) is slightly more complicated because we can approach the definition of this measure in a number of ways. For example, the definition in Delgado and Tibau [22] is essentially the same as the expression given in Jurman et al. [64].[5] In the latter

---

[5]Jurman et al. [64] also gave a complicated expression for *mcc* in terms of counts in a multiclass confusion matrix.

reference, the authors also presented a formula that can be re-expressed as

$$mcc = \frac{\sum_{j=1}^{k} \sum_{i=1}^{n} (X_{ij} - \overline{X_j})(Y_{ij} - \overline{Y_j})}{\sqrt{\sum_{j=1}^{k} \sum_{i=1}^{n} (X_{ij} - \overline{X_j})^2 \times \sum_{j=1}^{k} \sum_{i=1}^{n} (Y_{ij} - \overline{Y_j})^2}}. \tag{5.2}$$

In the above formula, $\overline{X_j}$ and $\overline{Y_j}$ are respectively the means of the j-th columns of indicator matrices $\boldsymbol{X} = [X_{ij}]_{n \times k}$ and $\boldsymbol{Y} = [Y_{ij}]_{n \times k}$ (these matrices were used earlier to demonstrate an alternative computation of confusion matrices). Equation (5.2) is equivalent to

$$mcc = \frac{n \times \sum_{i=1}^{k} n_{ii} - \sum_{i=1}^{k}(n_{i.} \times n_{.i})}{\sqrt{(n^2 - \sum_{i=1}^{k} n_{i.}^2) \times (n^2 - \sum_{i=1}^{k} n_{.i}^2)}}. \tag{5.3}$$

This is essentially the same as formula (25) in Grandini et al. [43].

The multiclass Matthews correlation coefficient is superior to *accuracy* as a measure of overall performance [14]. Delgado and Tibau [22] showed that *mcc* coincides with *kappa* when the confusion matrix is symmetric but their behaviour can diverge to the point that *kappa* should be avoided in favour of the more robust *mcc* when comparing classifiers. In particular, they noted that *kappa* is inadequate when there is imbalance in the marginal distribution of the classes.

```
# Compute mcc for the ML Classfier

# Compute mcc Using Definition
(sum_nii <- ML_pv_cm %>% pluck(1) %>% diag() %>% sum())
## [1] 41
ni.; n.i; n # from kappa calculation example
##    normal chemical    overt
##        21       12       11
##    normal chemical    overt
##        23       11       10
## [1] 44
(n*sum_nii - sum(ni.*n.i)) /
  sqrt((n^2-sum(ni.^2))*(n^2-sum(n.i^2)))
## [1] 0.893

# Compute mcc Using the yardstick Package
ML_pv %$% mcc_vec(group, pred_class)
## [1] 0.893
```

In practice, it is good to have some idea of the extent of class imbalance, if any. For the diabetes classification problem, you can get some idea from the pie chart that was obtained earlier during data exploration. You can also proceed as demonstrated below.

```
Test_data %$% table(group) %>% prop.table()
##    normal chemical     overt
##     0.523    0.250     0.227
```

The distribution of **group** shows that there is some imbalance, but it is not to a significant degree.

The summary below shows that classification accuracy for the ML classifier is very high. The remaining two measures are on the high end and they are quite close to one another (this is not surprising since the confusion matrix for the ML classifier is nearly symmetric).

```
# Overall Measures for the ML Classifier

ML_pv_cm %>% summary() %>% slice(1,2,7)
## # A tibble: 3 x 3
##    .metric    .estimator   .estimate
##    <chr>      <chr>             <dbl>
## 1 accuracy   multiclass        0.932
## 2 kap        multiclass        0.891
## 3 mcc        multiclass        0.893
```

In light of the above overall measures and the macro/micro averaged estimates obtained in the last section, it appears that the ML classifier has done a good job at solving the diabetes classification problem (confirming the initial impression obtained earlier from the confusion matrix in Figure 5.5). In the next section, we see whether further support for this assessment can be found when we examine some relevant performance curves and ranking measures derived from them.

5.3 Performance Curves and Surfaces

ROC analysis of a k-class problem with $k > 2$ can be challenging especially when you have to work with ROC hypersurfaces and the volume under such surfaces; see Wandishin and Mullen [112] to get a feel of what is involved for the case $k = 3$. The problems you face range from the apparently simple task of coordinate axes specification to the practical problem of visualizing the hypersurface. A simple expedient that is used in practice to overcome these problems makes use of the class reference formulation [35]. This requires you to take an OvR perspective; see Figure 1.7, for example.

Some similar concerns arise when you consider use of volume under the surface (VUS) for multiclass ROC analysis. Practical methods to determine VUS are somewhat limited. The numerical integration approach proposed by Landgrebe and Duin [70] works well only for $3 \leq k \leq 6$. A probabilistic approach with nonparametric estimation may also be used to obtain VUS. Kapasný and Řezáč [65] provide an example of this alternative for the 3-class problem; see Liu et al. [73] for a more comprehensive discussion.

Simpler alternatives are available for multiclass ROC analysis that do not require use of ROC hypersurfaces and determination of VUS. This requires you to settle for multiclass ROC curves and AUC, and use techniques like those discussed in Fawcett [35] and Hand [56], for example.

5.3.1   Multiclass ROC

There are several issues you need to address when you attempt to use hyper-surfaces in the ROC analysis of multiclass problems. The curse of dimension-ality is clearly one of them since visualization of the complete hypersurface is impossible when there are more than three classes. Another more basic issue is the question of what variables to use for the axes of the coordinate system when you want to define an ROC hypersurface.

To see what is involved in the axes-specification issue, consider again the ROC curve for binary classifiers. Typically, such curves are plots of tpr versus fpr with tpr on the vertical axis as seen in Figure 3.2, for example. This is not the only way to display a binary ROC curve. There is no loss of information if instead, you use fpr on the vertical axis. Other equivalent (in the sense of information content) graphical representations are possible.[6] For example, the inverse ROC curve in Ferri et al. [38] is a plot of fnr versus fpr. You can also use the true rates to define the axes and obtain a plot of tpr versus tnr. Figure 5.7 is an example of such a plot; it is an alternative that is equivalent to Figure 3.2.

When working with ROC hypersurfaces for multiclass classifiers, it is con-venient for the coordinate axes to be defined by the true rates for the various classes; see Figure 1 of Li and Fine [72], for example. However, as noted by Wandishin and Mullen [112], this simplification comes at the cost of lost infor-mation, specifically, all information about misclassification errors. They also discussed the broader approach introduced by Scurfield [100] that involves consideration of six ROC surfaces for a 3-class problem.

When the reference class is clearly defined for a k-class problem, one way to include error rates in the definition of the ROC hypersurface is to allow the true rate for the reference class to define one axis and use false rates that involve this class to define the remaining $k - 1$ axes, e.g., for a 3-class problem, if class 1 is the reference class, then the ROC surface is defined by the true rate tr_1 and false rates fr_{12} and fr_{13} (subscripts refer to the applicable classes). This approach may be viewed as a natural generalization of that used for the 2-class problem. In this case, an implicit assumption is that the true rates of

---

[6]This is similar to what we noted earlier about different possible representations of a confusion matrix.

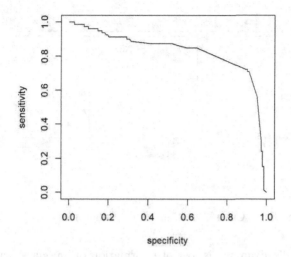

FIGURE 5.7
Plot Equivalent to the NN Classifier ROC Curve in Chapter 3

the remaining classes and false rates involving them are not as important for the problem at hand.

Given the limited utility of ROC hypersurfaces (at least for now), we will not pursue them further here (although we'll briefly revisit their use when discussing multiclass AUC). Instead, consider the popular alternative approach to multiclass ROC analysis based on an OvR collection of binary ROC curves. Recall that this approach is referred to as the class reference formulation by Fawcett [34]. The curves in such a collection may be displayed in a single plot like that shown in Figure 1.7 or separately as shown in Figure 5.8.

```
# Class Reference ROC Curves for the ML Classifier

# ROC Curves Using the yardstick Package
library(yardstick)
ML_pv %>%
  roc_curve(group, prob_normal:prob_overt) %>% autoplot()
```

The ROC curves in Figure 5.8 suggest excellent performance by the ML classifier. It is certainly necessary but not sufficient for each binary classifier in the OvR collection to have excellent performance in order for the same to be said about the multiclass classifier. This is what we see in the figure. Calculating the multiclass AUC should provide further insight. This will be taken up later in the next section.

Advantages of the class reference approach include linear complexity (i.e., it increases linearly with the number of classes) and easy visualization.

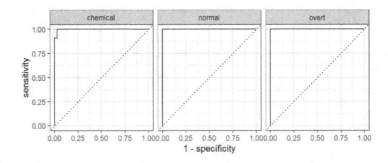

FIGURE 5.8

Class Reference ROC Curves for the ML Classifier

However, one disadvantage is loss of information on misclassification errors. These points were noted by Wandishin and Mullen [112] among others. We'll mention another disadvantage in the next section.

5.3.2 Multiclass AUC

Using the approach based on volume under the surface (VUS) to multiclass ROC analysis is an extension to what is involved when similar analysis is done for binary classifiers, but it suffers greatly from both computational complexity and interpretability [66]. There is also the concern that VUS may not be a very useful performance measure since the VUS of a random classifier approaches that of a perfect classifier as k increases [27]. In any case, for k-class problems with $k > 2$, the numerical integration approach proposed by Landgrebe and Duin [70] may be used to obtain the volume under the ROC hypersurface (VUS). As noted by these authors, the method works well for $3 \leq k \leq 6$, but for $k > 6$, they suggested that the only viable alternative is to use the Hand and Till [56] methodology (we discuss this later).

A probabilistic approach may also be applied for VUS estimation. For example, consider the 3-class problem in Wandishin and Mullen [112]. Their decision criteria for the problem is equivalent to the prediction function

$$\hat{f}(\boldsymbol{x}) = 1 \times I(S(\boldsymbol{x}) \leq c_1) + 2 \times I(c_1 < S(\boldsymbol{x}) \leq c_2) + 3 \times I(S(\boldsymbol{x}) > c_2),$$

where $S(\boldsymbol{x})$ is the score of a case with feature vector $\boldsymbol{x}$, c_1 and c_2 are specified thresholds, and $I(\cdot)$ is the indicator function. Varying the thresholds results in the ROC surface given in Figure 3 of their paper (the axes for their plot are determined by the true rates). The volume under this surface is given by

$$VUS = P(S_1 < S_2 < S_3), \tag{5.4}$$

where, for $i = 1, 2, 3$, S_i is the score for a random case from the i-th class.

A nonparametric estimator of the right-hand side of (5.4) is

$$\widehat{VUS} = \frac{1}{n_1 n_2 n_3} \sum_{i=1}^{n_1} \sum_{j=1}^{n_2} \sum_{k=1}^{n_3} I(S_{1i} < S_{2j} < S_{3k}).$$

where n_i is the number of scores from the i-th class. Note that (5.4) assumes the S_i's are *continuous* random variables; see Liu et al. [73] for a more general expression. Next, we use some synthetic data to illustrate calculation of $\widehat{VUS}$.

```
# Nonparametric Estimate of VUS

# Simulated Scores for 3 Classes
S1 <- c(-2, -0.11, -1.86); names(S1) = rep("A", 3)
S2 <- c(0.01, -0.45, 0.38, -0.69); names(S2) = rep("B", 4)
S3 <- c(0.12, 1.9, 0.5); names(S3) = rep("C", 3)

# Triple Sum Using for Loops
triple_sum <- 0
for (i in 1:3) {
  for (j in 1:4) {
    for (k in 1:3)
      if (S1[i] < S2[j] & S2[j] < S3[k]) {
        triple_sum <- triple_sum + 1
      }
  }
}
# Estimate of VUS
triple_sum / (3*4*3)
## [1] 0.75
```

A simpler approach to multiclass AUC is the one taken by Fawcett [35]. This involves calculation of a macro-weighted average of the AUCs from the OvR collection of binary ROC curves that results when you adopt a class reference formulation. However, as noted by Wandishin and Mullen [112], this approach is sensitive to changes in class priors. This is not the only problem with it. We'll return to this issue at the end of this section.

```
# Multiclass AUC by Macro-Averaging

# Using the yardstick Package
ML_pv %>% roc_auc(group, prob_normal:prob_overt,
    estimator = "macro_weighted")
## # A tibble: 1 x 3
##   .metric  .estimator     .estimate
##   <chr>    <chr>              <dbl>
## 1 roc_auc  macro_weighted     0.998
```

The preceding calculation of multiclass AUC by macro-averaging simply uses the `roc_auc` function from the **yardstick** package. Although longer (and it is something you'll probably not do in practice), the next approach to the problem is quite instructive since we'll use it later to obtain a multiclass estimate of the H-measure.

To begin, we write a function called `OvR_pv_fn()` to obtain the tibble containing predicted probabilities and actual classes for the i-th member in the OvR collection of tibbles. We then use this function in the `OvR_AUC_fn()` function that allows us to obtain the i-th AUC in the OvR collection of AUCs.

```
# Multiclass AUC by Macro-Averaging (cont'd)

# Levels of group
classes # this variable was created earlier
## [1] "normal"   "chemical" "overt"

# Function for i-th tibble in OvR Collection of tibbles
OvR_pv_fn <- function(i) {
  ML_pv[,c(i,5)] %>%
    mutate(group = fct_other(group, keep = classes[i])) %>%
    set_names(c("prob_Yes", "group"))
}

# Function for i-th AUC in OvR Collection of AUCs
OvR_AUC_fn <- function(i) {
  OvR_pv_fn(i) %>% roc_auc(group, prob_Yes) %$% .estimate
}

# Obtain the Multiclass AUC
(OvR_AUCs <- 1:3 %>% map(~ OvR_AUC_fn(.)) %>% unlist())
## [1] 1.000 0.992 1.000
(wts <- ML_pv %>% count(group) %>% mutate(p = n/sum(n)) %$% p)
## [1] 0.523 0.250 0.227
(wts * OvR_AUCs) %>% sum() # macro-weighted AUC
## [1] 0.998
```

Note that the OvR collection of ROC curves that we obtained earlier with the `roc_auc()` function may also be produced with help from the `OvR_pv_fn()` function, and the `grid.arrange()` function from the **gridExtra** package. We leave this as an exercise for the reader.

In practice, a more common approach to multiclass AUC is to use the M-measure proposed by Hand and Till [56]. This measure is calculated by default when you use the `roc_auc()` function from **yardstick**. This is illustrated next for the ML classifier.

```
# Multiclass AUC for the ML Classifier

ML_pv %>% roc_auc(group, prob_normal:prob_overt)
## # A tibble: 1 x 3
##   .metric .estimator .estimate
##   <chr>   <chr>          <dbl>
## 1 roc_auc hand_till      0.999
```

The multiclass AUC that we obtain suggests excellent ability by the ML classifier to separate the three classes. The result we obtain when you combine this observation with what we saw earlier for the class reference ROC curves confirms the excellent performance noted earlier for this classifier.

Next, we provide some additional details on the Hand and Till [56] approach to multiclass AUC. To begin, note that their approach generalizes the estimate of AUC for a binary classifier based on (3.3) in Chapter 3. More precisely, their measure is an estimate of the overall ability of a multiclass classifier to separate k classes that is given by

$$M = \frac{2}{k(k-1)} \sum_{i<j} \hat{A}_{ij}. \tag{5.5}$$

where

$$\hat{A}_{ij} = \frac{\hat{A}_{i|j} + \hat{A}_{j|i}}{2}$$

is an estimate of the measure of separability between class i and j. Here, $\hat{A}_{i|j}$ and $\hat{A}_{j|i}$ are estimates of $P(S_{i|i} > S_{i|j})$ and $P(S_{j|j} > S_{j|i})$, respectively, where an expression like $S_{r|s}$ denotes the class r membership score for a randomly selected case from class s (we implicitly assume that cases are assigned to class r when such scores exceed a pre-specified threshold).

```
# Compute Hand & Till (2001) M-Measure

pv_tb <-
  tibble(
    A = c(0.501,0.505,0.341,0.168,0.178,0.22,0.22,0.207,
          0.338,0.355),
    B = c(0.228,0.251,0.408,0.458,0.474,0.362,0.489,0.511,
          0.17,0.204),
    C = c(0.271,0.244,0.251,0.374,0.348,0.418,0.291,0.282,
          0.492,0.441),
    group = c("A","A","A","B","B","B","B","C","C","C") %>%
      as.factor()
  )
```

We use synthetic data in the `pv_tb` tibble to illustrate what is involved in calculation of the M-measure; we'll use this term when referring to the measure defined by (5.5).[7] Note that in the following manual calculation of the measure, we adapted the formula given by (3.3) to obtain the estimates $\hat{A}_{1|2}$ and $\hat{A}_{2|1}$.

```
# Compute Hand & Till (2001) M-Measure (cont'd)

# Obtain AB_tb to Compute Ahat_1|2 and Ahat_2|1
(AB_tb <- pv_tb %>% slice(1:7) %>% select(-C) %>%
  mutate(rankA = rank(A), rankB = rank(B)))
## # A tibble: 7 x 5
##       A     B group rankA rankB
##   <dbl> <dbl> <fct> <dbl> <dbl>
## 1 0.501 0.228 A         6     1
## 2 0.505 0.251 A         7     2
## 3 0.341 0.408 A         5     4
## 4 0.168 0.458 B         1     5
## 5 0.178 0.474 B         2     6
## 6 0.22  0.362 B       3.5     3
## 7 0.22  0.489 B       3.5     7

# Compute Ahat_1|2
n1 <- 3; n2 <- 4
S <- AB_tb %>% filter(group == "A") %$% sum(rankA)
(S - n1*(n1+1)/2) / (n1*n2)
## [1] 1

# Compute Ahat_2|1
n1 <- 4; n2 <- 3
S <- AB_tb %>% filter(group == "B") %$% sum(rankB)
(S - n1*(n1+1)/2) / (n1*n2)
## [1] 0.917
```

The above calculations show that $\hat{A}_{1|2} = 1$ and $\hat{A}_{2|1} = 0.917$ (we refer to class "A", "B" and "C" as class 1, 2, and 3, respectively). Similar calculations show that $\hat{A}_{1|3} = 0.889$, $\hat{A}_{3|1} = 1$, $\hat{A}_{2|3} = 0.667$ and $\hat{A}_{3|2} = 0.667$. Hence,

$$M = \frac{2}{3(3-1)} \left(\frac{1 + 0.917}{2} + \frac{0.889 + 1}{2} + \frac{0.667 + 0.667}{2} \right) = 0.857.$$

The value you get when you use the `roc_auc()` function is 0.856.

Kleiman and Page [66] noted some issues with M-measure. In particular, they observed that this measure can fail to return the ideal value of 1, even

---

[7]Note that we used class membership scores in our discussion of this measure instead of estimated class membership probabilities as was done by Hand and Till [56]. As noted by the authors, the use of scores in the formulation is also valid.

when for every case, a classifier gives the correct class the highest probability. They also noted that the measure is not the probability that randomly selected cases will be ranked correctly. As an alternative, they proposed their AUC_μ measure. This measure was motivated by the need to deal with computational complexity and the issue of what properties of AUC for binary classifiers are worth preserving. It is based on the relationship between AUC and the Mann Whitney U-statistic (generalizing this statistic was a key step in development of AUC_μ). The authors claim that AUC_μ is a fast, reliable, and easy-to-interpret method for assessing the performance of a multiclass classifier. However, despite this claim, their measure has yet to gain wide acceptance, perhaps due to the fact that the measure was only recently proposed.

Perhaps a more problematic issue is the one that arises due to incoherence of AUC when used as a performance measure for binary classifiers; our earlier discussion in Chapter 3 showed why this is a concerning issue. Binary AUCs are involved (whether directly or indirectly) in the calculation of macro-weighted AUC, M-measure, and the AUC_μ measure. It stands to reason that the incoherence of binary AUCs leads to similar issues with these multiclass AUC estimates. We need a better understanding of how the incoherence of binary AUC leads to problems with the multiclass analogs considered so far. This is possibly, if not likely, an area of current research into multiclass AUC. In the meantime, we can consider the use of a macro-weighted H-measure. This is demonstrated below for the ML classifier.

```
# Calculate a Multiclass H-Measure

OvR_HM_fn <- function(i) {
  OvR_pv_fn(i) %>%
    mutate(group = fct_recode(group, "Yes" = classes[i],
      "No" = "Other")) %$% HM_fn(group, prob_Yes)
}
(OvR_HMs <- 1:3 %>% map(~ OvR_HM_fn(.)) %>% unlist())
## [1] 1.000 0.909 1.000
(wts * OvR_HMs) %>% sum() # macro-weighted H-Measure
## [1] 0.977
```

The above value for macro-weighted H-measure supports the earlier evidence provided for the excellent performance of the ML classifier.

5.4   Inferences for Performance Parameters

As in Chapter 2, let $\widehat{Y}$ and Y denote the predicted and actual target (i.e., response) variables, respectively, for a randomly selected case. For a k-class

TABLE 5.2
Joint PMF of $\widehat{Y}$ and Y

	Actual			
Predicted	$Y = 1$	$Y = 2$	$\cdots$	$Y = k$
$\widehat{Y} = 1$	p_{11}	p_{12}	$\cdots$	p_{1k}
$\widehat{Y} = 2$	p_{21}	p_{22}	$\cdots$	p_{2k}
$\vdots$	$\vdots$	$\vdots$	$\vdots$	$\vdots$
$\widehat{Y} = k$	p_{k1}	p_{k2}	$\cdots$	p_{kk}

problem, the joint probability mass function (PMF) of these random variables is given in Table 5.2 where

$$p_{ij} = P(\widehat{Y} = i, Y = j), \quad i, j = 1, \ldots, k.$$

Given the joint distribution in Table 5.2, the marginal distribution of $\widehat{Y}$ is given by

$$P(\widehat{Y} = i) = \sum_{j=1}^{k} p_{ij} = p_{i\cdot}, \quad i = 1, \ldots, k,$$

and that for Y is given by

$$P(Y = j) = \sum_{i=1}^{k} p_{ij} = p_{\cdot j}, \quad j = 1, \ldots, k.$$

The joint probabilities p_{ij} are the parameters that define the joint distribution in Table 5.2. A number of performance parameters (e.g., *Accuracy* and *Kappa*) for the multiclass problem are functions of these parameters. In this section, we consider some inferential procedures for these parameters.

5.4.1   Multiclass Performance Parameters

We can define some multiclass performance parameters in terms of the probabilities in Table 5.2. For example, the multiclass *Kappa* parameter is defined by

$$Kappa = \frac{(A - E)}{(1 - E)}, \tag{5.6}$$

where A is equal to the *Accuracy* parameter

$$Accuracy = P(\widehat{Y} = Y) = \sum_{i=1}^{k} P(\widehat{Y} = i, Y = i) = \sum_{i=1}^{k} p_{ii}, \tag{5.7}$$

which simplifies to

$$E = \sum_{i=1}^{k} P(\widehat{Y} = i) P(Y = i) = \sum_{i=1}^{k} p_{i\cdot} p_{\cdot j}$$

when $\widehat{Y}$ and Y are statistically independent.

Next, for $i = 1, \ldots, k$, let

$$TPR_i = P(\widehat{Y} = i \mid Y = i) = \frac{p_{ii}}{p_{.i}}$$

and

$$PPV_i = P(Y = i \mid \widehat{Y} = i) = \frac{p_{ii}}{p_{i.}}.$$

The macro-averaged multiclass *Sensitivity* (or, *Recall*) parameter is obtained by averaging over classes, i.e.,

$$TPR = \frac{1}{k} \sum_{i=1}^{k} TPR_i = \frac{1}{k} \sum_{i=1}^{k} \frac{p_{ii}}{p_{.i}}. \tag{5.8}$$

Similarly, the macro-averaged multiclass *PPV* (or, *Precision*) parameter is defined by

$$PPV = \frac{1}{k} \sum_{i=1}^{k} PPV_i = \frac{1}{k} \sum_{i=1}^{k} \frac{p_{ii}}{p_{i.}}. \tag{5.9}$$

Averaging the harmonic mean of TPR_i and PPV_i over classes yields the following macro-averaged multiclass parameter

$$F_1\text{-Measure} = \frac{1}{k} \sum_{i=1}^{k} \left(2 \times \frac{TPR_i \times PPV_i}{TPR_i + PPV_i} \right) = \frac{1}{k} \sum_{i=1}^{k} \frac{2p_{ii}}{p_{i.} + p_{.i}}. \tag{5.10}$$

There is a less commonly used alternative to (5.10). This is simply the harmonic mean of (5.8) and (5.9). The corresponding estimate is a special case of the $Fscore_M$ given in Sokolova and Lapalme [103].

There is also a popular micro-averaged multiclass parameter that is obtained by taking the harmonic mean of the corresponding micro-averaged multiclass *Precision* and *Recall* parameter. As noted in Takahashi et al. [106], the resulting harmonic mean simplifies to the right-hand side of (5.7).

5.4.2 MLE of Performance Parameters

When you assume a multinomial distribution for the cell counts in a confusion matrix, you can obtain maximum likelihood estimators of the p_{ij}'s in Table 5.2. The MLEs are given by [106]

$$\hat{p}_{ij} = \frac{n_{ij}}{n}, \quad i, j = 1, \ldots, k,$$

where n_{ij} are the counts in Table 1.6, and n is the sum of the n_{ij}'s. By the invariance property of MLEs, you can obtain ML estimates of the various performance parameters by substituting p_{ij}'s in the formula for a multiclass parameter by the corresponding $\hat{p}_{ij}$'s where applicable. For example, the MLE of *Kappa* is $(\hat{A} - \hat{E})/(1 - \hat{E})$ where

$$\hat{A} = \sum_{i=1}^{k} \frac{n_{ii}}{n} \text{ and } \hat{E} = \sum_{i=1}^{k} \left(\sum_{j=1}^{k} \frac{n_{ij}}{n} \right) \left(\sum_{j=1}^{k} \frac{n_{ji}}{n} \right).$$

Note the $\hat{A}$ is the MLE of the *Accuracy* parameter.

```
# MLEs of Kappa and F1-Measure

# MLEs of Cell Probabilities
(pij <- ML_pv_cm %>% pluck(1) %>% prop.table())

# MLE of Kappa Parameter for the ML Classifier
pii <- pij %>% diag()
pi. <- pij %>% rowSums()
p.i <- pij %>% colSums()
Ahat <- sum(pii); Ehat <- sum(pi. * p.i)
(Ahat - Ehat) / (1 - Ehat)
## [1] 0.891

# MLE of Macro F1-Measure for the ML Classifier
(2/3) * sum(pii / (pi. + p.i))
## [1] 0.925

# MLE of Micro F1-Measure for the ML Classifier
sum(pii) # see Takahashi et al (2022, p. 4963)
## [1] 0.932

# Check: Using the yardstick Package
# ML_pv %$% kap_vec(group, pred_class)
# ML_pv %$% f_meas_vec(group, pred_class, estimator = "macro")
# ML_pv %$% f_meas_vec(group, pred_class, estimator = "micro")
```

The commented portion at the bottom of the above code segment shows how to use the **yardstick** package to obtain the required MLEs.

5.4.3   Interval Estimation of Performance Parameters

The confidence interval for the *Accuracy* parameter that was given in Section 2.5.1 for binary classifiers is also applicable to the corresponding multiclass parameter (5.7). It suffices to note that now, x_0 is the sum of counts on the diagonal of the multiclass confusion matrix.

```
# 95% Confidence Interval for Accuracy Parameter

n <- ML_pv %>% nrow()
x0 <- ML_pv %$% table(pred_class, group) %>% diag() %>% sum()
alpha <- 0.05
theta_L <- qbeta(alpha/2, x0, n - x0 + 1)
theta_U <- qbeta(1 - alpha/2, x0 + 1, n - x0)
c(theta_L, theta_U)
## [1] 0.813 0.986

# Check: Using the caret package
# ML_pv %$% confusionMatrix(pred_class, group)
```

Takahashi et al. [106] considered the interval estimation problem for three multiclass F_1-measure parameters. Their confidence intervals were based on the asymptotic normal distribution of the maximum likelihood (ML) estimators of the p_{ij}'s in Table 5.2, and use of the multivariate delta-method [71, p. 315]. They gave the following limits

$$\sum_{i=1}^{k} \hat{p}_{ii} \pm z_{\alpha/2} \sqrt{\frac{1}{n} \left(\sum_{i=1}^{k} \hat{p}_{ii} \right) \left(1 - \sum_{i=1}^{k} \hat{p}_{ii} \right)}$$

for a large sample $100(1 - \alpha)\%$ confidence interval for micro-average F_1-measure.

```
# 95% Confidence Interval for Micro F1-Measure

n <- ML_pv %>% nrow()
pij <- ML_pv %$% table(pred_class, group) %>% prop.table()
miF1 <- pij %>% diag() %>% sum() # micro average F1 score
miF1 + 1.96 * c(-1, 1) * sqrt(miF1*(1-miF1)/n)
## [1] 0.857 1.006
```

Instead of using their large sample method to obtain a confidence interval for the macro-average F_1-measure metric, you can consider the use of a bootstrap approach [20]. Let

- θ denote the RHS of (5.10) for the NN classifier,

- $\hat{\theta}$ denote a point estimate (e.g., MLE) of θ,

- $\hat{\theta}_1^*, \ldots, \hat{\theta}_B^*$ denote B bootstrap replicates of $\hat{\theta}$,

- q_L and q_U denote the lower and upper $\alpha/2$-quantiles obtained from the $\hat{\theta}_i^*$'s.

Two $100(1 - \alpha)\%$ bootstrap confidence intervals for θ are $(2\hat{\theta} - q_U, 2\hat{\theta} - q_L)$ and (q_L, q_U). The first is called the basic bootstrap interval and the second is the percentile interval.

The boot.ci() function in the **package** may be used to obtain the above-mentioned bootstrap confidence intervals. Instead of doing this, we use the boot() function in this package to obtain the bootstrap replicates for the macro F_1-*score* in the following code segment, and use the replicates to obtain estimates of q_L and q_U. These estimates are then used together with the corresponding MLE to obtain a basic bootstrap confidence interval for macro F_1-measure.

```
# 95% Bootstrap CI for Macro-Average F1-Measure

# Function to Compute the Performance Parameter
pm_fn <- function(dat, ind) {
  pred <- dat[ind, 1] %>% as.factor()
  act <- dat[ind, 2] %>% as.factor()
  f_meas_vec(act, pred, estimator = "macro")
}

# Obtain Bootstrap Replicates for Macro F1-Measure
library(boot)
set.seed(11423)
boot_obj <- ML_pv %$% cbind(pred_class, group) %>%
  boot(statistic = pm_fn, R = 2000)
replicates_tb <- tibble(replicates = boot_obj$t[, 1])

# Percentile Bootstrap Confidence Interval for Macro F1-Measure
(q_lu <- replicates_tb %$% quantile(replicates,c(0.025, 0.975)))
##   2.5% 97.5%
## 0.822 1.000

# MLE of Macro F1-Measure and its Std Error
(theta_hat <- ML_pv %$%
  f_meas_vec(group, pred_class, estimator = "macro"))
## [1] 0.925
replicates_tb %$% sd(replicates)
## [1] 0.0448

# Basic Bootstrap Confidence Interval for Macro F1-Measure
2*theta_hat - rev(q_lu)
## 97.5%  2.5%
## 0.851 1.029
```

You get slightly different results when you use boot.ci() function because the **boot** package uses its own quantile function. Note that this function also allows you to obtain bias-corrected and accelerated (BCa) bootstrap intervals. Such intervals have better performance [23, 96]. The second reference has an **R** function that you can use to compute BCa intervals.

The bootstrap confidence interval approach is quite versatile. You can also use it to obtain confidence intervals for other multiclass performance parameters like (5.8) and (5.9). It suffices to modify the last line of the pm_fn() function in the above code segment.

5.5 Exercises

1. Construct a decision tree classifier for the diabetes classification problem. Use the same training and test datasets (i.e., Train_data and Test_data) as that used for the ML classifier given by (5.1).

   (a) Obtain a display of the decision tree.

   (b) Obtain the confusion matrix and threshold measures.

   (c) Obtain the class reference ROC curves and multiclass AUC.

2. (a) Use the sens() function in the **yardstick** package to verify the multiclass *sensitivity* estimates given for the ML classifier.

   (b) Estimate multiclass *specificity* for the ML classifier without using **yardstick**.

3. Consider the following confusion matrix reported in Zhang et al. [118] for a 5-class classifier.

```
##                   predicted class label
## true class label   0    1    2    3    4
##                0  145    1    2    1    0
##                1    5  256   22    9    6
##                2    5   24  234   36   19
##                3    1   18   32  243   25
##                4    1    5    9   38  254
```

   (a) Use the **yardstick** package to obtain macro average estimates of *precision, recall* and F_1-measure.

   (b) Obtain the **OvR** collection of key counts and use it to obtain micro average estimates of the measures mentioned in part (a).

4. Dong et al. [25] proposed a new approach based on a Mixed Neural Network (MNN) to classify sleep into five stages with one awake stage (W), three sleep stages (N1, N2, N3), and one rapid eye movement stage (REM). The confusion matrix they obtained in their study for this classifier is given below.

```
# Confusion Matrix of MNN Classifier

##              Actual
## Predicted    W      N1     N2    N3    REM
##      W     5022    407    130    13    103
##      N1    577    2468    630     0    258
##      N2    188     989  27254  1236    609
##      N3     19       4   1021  6399      0
##      REM   395     965    763     5   9611
```

(a) Consider the parameters that are defined by (5.8) and (5.9). Obtain the MLE of these parameters.

(b) Compute the harmonic mean of the estimates obtained in part (a). Comment on what you have estimated.

(c) Obtain a 95% confidence interval for the metric in part (b).

5. Do the following comparisons between the ML classifier given by (5.1) and the DT classifier that you obtained in your answer to Exercise 1.

   (a) Compare (i) the confusion matrices, and (ii) overall and composite performance measures.

   (b) Compare the Hand and Till [56] AUCs.

   (c) Comment on the results you obtained for the comparisons.

6. The tibble given below contains point and interval estimation results from Takahashi et al. [106] for the F_1-measure that were based on maximum likelihood estimation and macro averaging. These results are part of what the authors obtained for the temporal sleep stage classification problem discussed in Dong et al. [25].

```
## # A tibble: 4 x 4
##    classifier estimate lower_limit upper_limit
##    <chr>        <dbl>      <dbl>       <dbl>
## 1 MNN          0.805      0.801       0.809
## 2 SVM          0.75       0.746       0.754
## 3 RF           0.724      0.72        0.729
## 4 MLP          0.772      0.768       0.776
```

(a) Use the `tibble()` function to reproduce the above results.

(b) Obtain a visual display of the confidence intervals.

(c) Comment on the relative performance of the classifiers.

6

Additional Topics in CPA

Knowing the material we have covered in the last five chapters is necessary but not sufficient if you want to analyze performance of classification algorithms. Whether you are involved with binary or multiclass CPA, the topics we'll discuss in this chapter are relevant because they help you deal with questions like the following.[1]

- What is data leakage and how can it impact performance analysis of classifiers?

- How can you detect and prevent overfitting of the underlying model that your classifier is based on?

- What is feature engineering and why is it important?

- What is the bias-variance trade-off in the context of predictive modeling?

- How can you properly evaluate your classifier without using the test dataset?

- How useful are resampling techniques for CPA and how can they be used for this purpose?

- What is hyperparameter tuning for a classifier and how does one perform it?

- Why is class imbalance an important issue when classifiers are trained and evaluated?

- How do we deal with the class imbalance problem in practice?

If you are new to CPA, the basic issues underlying some of the above questions are probably not the ones that comes to mind initially. This is not surprising because, unlike obvious concerns with accuracy and classification errors, not all of these questions deal directly with performance *measurement* even though their relevance to classifier performance is unquestionable. However, with further exposure to CPA, you will encounter these issues and it is a matter of time before you have to address them. For instance, this might

---

[1]This list is not exhaustive (e.g., we omitted questions about resampling techniques for comparing classifiers.) but we will focus on the listed questions in this chapter.

DOI: 10.1201/9781003518679-6

happen when you attempt to examine the reasons why a particular classification algorithm is underperforming for your problem (the cause could be due to overfitting or class imbalance or both, for example).

Thus, in this section, we will address the questions that were raised since they have relevance to classifier performance. Some topics like cross-validation have *direct* relevance to CPA, while others like data or classifier level approaches to deal with class imbalance do not have the same degree of relevance but are nonetheless important if you are interested in classifier performance.

6.1   Some Important Modeling Issues

The first four questions raised at the beginning of this chapter are concerned with modeling issues. To get some sense of the importance of a topic like overfitting, it suffices to note, for example, that Provost and Fawcett [90] devoted an entire chapter of their book to this topic.

Data leakage is a subtle problem but unless you know the effects of this issue, you can potentially run into problems when you apply an inappropriate approach to data preprocessing when using resampling techniques like k-fold cross-validation in classifier training and evaluation.

Choosing the right set of predictors for your classifier is clearly necessary if you want it to perform well for your classification problem. Data availability aside, to determine this set, you need to pay attention to aspects like choice, number and representation of predictors, dimension reduction, missing values imputation, and other required preprocessing. These are feature engineering problems that you can address by designing a specification that incorporates the required steps to deal with data issues.

The bias-variance trade-off is also important but the relevance of this issue depends on the context in which it is under consideration. The context dictates the meaning you give to properties like bias and variance. Unfortunately, the distinction has not been made sufficiently clear in the literature. Hence, it is potentially a source of confusion for students and others who want to take into account the trade-off when they train classifiers and analyze their performance.

In this section, we provide some information on the abovementioned important issues. For further information, readers can refer to [6, 16, 69, 90], for example.

6.1.1   Data Leakage

This is a problem you face when you use data outside your training dataset to create a predictive model. The extraneous information can lead to your model learning patterns and rules that may not be present when you apply it to unseen data in your problem. In other words, you wind up with a model that cannot be used in production.

There is a good chance you have data leakage when the model you created yields doubtful optimistic results. One technique you can use to avoid it is what we have done in earlier chapters when we used data splitting to create separate datasets for training and testing. The second dataset is usually set aside for the sole purpose of evaluating the model you independently obtained from training data.

Data leakage can occur when you employ k-fold cross-validation (we'll discuss this further later) to evaluate your model if you do not carry out any required preprocessing correctly. It occurs when you preprocess the entire dataset before using cross-validation to evaluate the model. The correct procedure is to perform preprocessing within each fold; see Figure 3.10 in Boehmke and Greenwell [6, p. 69] for an illustration of what is involved.

6.1.2 Overfitting: Causes, Effects and Remedies

An important quality that we seek in a trained classifier is generalizability, i.e., the ability to perform well with unseen data. To achieve this quality, it is important that you do not overfit the underlying model that the classifier relies on.[2] Rather than discovering generalizable rules or patterns, overfitted models tend to memorize patterns in the training data. Consequently, they will perform well when evaluated with training data but exhibit large generalization error when evaluated with unseen test data. Tuning models to optimize performance can also lead to overfitting, especially when the results from tuning is influenced by unusual data characteristics. The problem can also arise due to data leakage.

Thus, one sign of overfitting is discrepancy between training and test performance of your model when it is the only model under consideration, e.g., when your model performs much better with training data than with test data. You can also have overfitting with the model under evaluation when there is a competing model with better test-set accuracy but poorer performance with training data [16]. In general, there is greater likelihood for the problem to manifest itself when you use overly complicated models.

Certain models are prone to overfitting, e.g., models like decision trees and k-nearest neighbors that are very flexible and adaptable in the patterns that they can fit. For example, consider what can happen when you use a decision tree to solve your classification problem. Overfitting can result if you use an unduly large and complicated tree (as measured by the number of nodes). When you examine performance of such a classifier as you vary the number of nodes, it is possible to find that both training-set accuracy and test-set accuracy increasing initially as you increase the number of nodes, but at some point the two measures diverge with the former increasing without bound

---

[2]You can also underfit a model by not having sufficient predictors or if the model is too simple, but this is "less frequently a problem than overfitting" [95, p. 67].

while the latter continues to decrease; see Provost and Fawcett [90, p. 117] for a figure illustrating this point and a more in-depth discussion of overfitting in tree induction.

How can you prevent overfitting? Zumel and Mount [119] suggest the use of simple models that generalize better whenever possible. This makes sense in light of the above discussion on how the problem can arise with tree induction. Other suggestions they gave include using techniques like regularization and bagging.

Regularization allows you to constrain estimated coefficients in your model thereby reducing variance and decreasing out-of-sample error [6]; this can be achieved, for example, by using a classifier based on the penalized logistic regression model (such models shrink coefficients of less beneficial predictors to zero by imposing suitable penalties). Boehmke and Greenwell [6] provide some discussion on how to implement regularization with linear models through the use of ridge penalty, lasso penalty, or elastic nets.

Bagging refers to bootstrap aggregation that underpins early attempts to create ensemble learners. A random forest classifier is an example of such a learner that accomplishes model averaging (a key aspect of ensemble learning) through bagging. This reduces variance and minimizes overfitting [6].

6.1.3   Feature Engineering and Preprocessing

To solve a classification problem empirically, you need a suitable prediction model for the categorical target variable in the problem. A key part of the model is the set of features (i.e., predictors) that serves as inputs for your prediction. For a good model, you need a good selection of features, suitably preprocessed if necessary. The right number of features also matters since too many (few) features in your model can lead to overfitting (underfitting).

Feature engineering is the process that you use to obtain a well-designed collection of features for your problem. Among other things, the process involves selection of a set of relevant features, application of suitable transformations and encodings of the variables to achieve more useful representations (some of which may be required by the classification algorithm in question), and other required preprocessing.

For numeric features, some of the procedures involved in feature engineering include dimension reduction, identification of near zero-variance predictors (i.e., those with a single unique value), removal of variables with large pairwise correlations, and normalization or standardization of the variables. To carry out these procedures, you add the required steps when applying the `recipe()` function from the **recipes** package by using helper functions like `step_pca()`, `step_zv()`, `step_corr()`, `step_range()` and `step_normalize()`.

To illustrate some transformations that are often performed during feature engineering, consider how to normalize or standardize numeric feature variables. Recall that normalization transforms numeric variables so that their range are restricted to the interval $[0, 1]$, and standardization transforms such variables so that they have zero means and unit standard deviations; see Exercise 1 for precise definitions.

```r
# Demonstrate Normalization of Features

# Create a Tibble with Three Numeric Variables from mtcars
mpg_tb <- mtcars %>% select(mpg, disp, wt) %>% tibble()
mpg_tb %>% print(n = 3)
## # A tibble: 32 x 3
##      mpg  disp    wt
##    <dbl> <dbl> <dbl>
## 1   21     160  2.62
## 2   21     160  2.88
## 3   22.8   108  2.32
## # ... with 29 more rows

# Function to Compute Relevant Statistics
stats_fn <- function(x){
  statistics <- c(mean(x), sd(x), min(x), max(x))
  names(statistics) <- c("Mean", "Std Dev", "Min", "Max")
  statistics %>% round(2)
}

# Statistics from Normalized Features
library(recipes) # a core package in tidymodels
recipe(mpg ~ disp + wt, data = mpg_tb) %>%
  step_range(all_numeric_predictors()) %>%
  prep() %>% juice() %>%
  apply(., 2, stats_fn)
##           disp   wt   mpg
## Mean      0.40 0.44 20.09
## Std Dev   0.31 0.25  6.03
## Min       0.00 0.00 10.40
## Max       1.00 1.00 33.90
```

Notice that the first argument in the call to the `recipe()` function is a model formula. This formula helps **R** distinguish between the response (i.e., target) and feature variables in the `mpg_tb` tibble. Since we only want to normalize the features, we see from the summary statistics that these variables are in the range we expect.

Next consider standardization of features in the `mpg_tb` tibble.[3] As shown in the following code segment, the means and standard deviations of these variables take on values that we expect.

---

[3] Note that the reference to standardization here is in line with how the term is commonly used in statistics. Also, note that the helper function in **recipes** to perform standardization uses the misleading name `step_normalize`.

```
# Demonstrate Standardization of Features

# Statistics from Standardized Features
recipe(mpg ~ disp + wt, data = mpg_tb) %>%
  step_normalize(all_numeric_predictors()) %>%
  prep() %>% juice() %>%
  apply(., 2, stats_fn)
##            disp    wt   mpg
## Mean       0.00  0.00 20.09
## Std Dev    1.00  1.00  6.03
## Min       -1.29 -1.74 10.40
## Max        1.95  2.26 33.90
```

You can, of course, normalize/standardize the feature variables and obtain the summary statistics without using functions in the **recipes** package (for statistical modeling, its use with other core packages in the **tidymodels** meta-package is recommended); see Exercise 1.

As another example of feature engineering, consider dummy and one-hot encoding of features. These encodings are important procedures that are often applied to categorical predictor variables as part of the modeling process. The former is required when such variables are features for linear models, and use of the latter can yield more interpretable splits when creating decision trees and rule-based models [68].

```
# Demonstrate Dummy and One-Hot Encoding

# Create a Tibble with Simulated Categorical Data
set.seed(170124)
wxy_tb <-
  tibble(
    w = factor(sample(c("F", "M"), 5, replace = TRUE)),
    x = factor(sample(LETTERS[1:3], 5, replace = TRUE)),
    y = factor(sample(c("N", "Y"), 5, replace = TRUE))
  )
wxy_tb
## # A tibble: 5 x 3
##   w     x     y
##   <fct> <fct> <fct>
## 1 F     C     Y
## 2 F     B     N
## 3 M     C     N
## 4 F     A     N
## 5 M     B     N
```

By design, we consider the two first variables in the wxy_tb tibble that was just created as the features involved in the dummy and one-hot encoding demonstration. In general, both encodings involve creation of a set of $n \times 1$ vectors containing values in $\{0, 1\}$ for each categorical feature, where n is the size of the dataset. For a variable with m categories, the first (second) type of encoding creates $m-1$ (m) such vectors. The vector for each category indicates whether it is present for cases in the dataset (the ones indicate presence). Note that the response variable in wxy_tb is not subjected to any encoding.

```
# Demonstrate Dummy and One-Hot Encoding (cont'd)

# Dummy Encoding
recipe(y ~ w + x, data = wxy_tb) %>%
  step_dummy(all_nominal_predictors()) %>%
  prep() %>% juice()
## # A tibble: 5 x 4
##    y        w_M   x_B   x_C
##    <fct> <dbl> <dbl> <dbl>
## 1 Y         0     0     1
## 2 N         0     1     0
## 3 N         1     0     1
## 4 N         0     0     0
## 5 N         1     1     0

# One-Hot Encoding
recipe(y ~ w + x, data = wxy_tb) %>%
  step_dummy(all_nominal_predictors(), one_hot = TRUE) %>%
  prep() %>% juice()
## # A tibble: 5 x 6
##    y        w_F   w_M   x_A   x_B   x_C
##    <fct> <dbl> <dbl> <dbl> <dbl> <dbl>
## 1 Y         1     0     0     0     1
## 2 N         1     0     0     1     0
## 3 N         0     1     0     0     1
## 4 N         1     0     1     0     0
## 5 N         0     1     0     1     0
```

The datasets that you encounter in practice often have missing values for some variables. Earlier, we used the replace_na() function from the **tidyr** package to impute approximations for these values when preparing data for the Titanic survival classification problem. However, you can also use the recipe() function to perform such imputations through the use of steps specified by helper functions like step_medianimpute(), step_knnimpute() or step_bagimpute(); see Boehmke and Greenwell [6, p. 50] for further discussion of these alternatives.

The discussion up to this point is only a brief account of what is possible with feature engineering. A more comprehensive coverage of this important topic may be found in Kuhn and Johnson [68]. A good introductory account may be found in Chapter 8 of Kuhn and Silge [69].

6.1.4 Trade-Off Between Bias and Variance

The bias-variance trade-off is a well-known issue in statistics, particularly in the context of point estimation. It is also an issue when we consider prediction problems in machine learning. The trade-off arises for certain problems when we consider the following decomposition (this follows immediately from the definition of variance of a random variable)

$$E(e^2) = var(e) + E(e)^2. \tag{6.1}$$

For point estimation,

$$e = \hat{\theta} - \theta$$

represents the estimation error for estimator $\hat{\theta}$ of parameter θ. The left-hand side of (6.1) represents mean squared error (MSE) for an estimator, and on the right-hand side, we have its variance and squared bias. Here, variance measures the expected squared deviation of an estimator from its mean, and bias is the expected error made by an estimator. Variance (bias) can be small but MSE can still be large if bias (variance) is large. This trade-off is relevant when your interest lies in the quality of an estimator of a performance parameter for your classification problem; see Cichosz [16] for a discussion of this aspect of the trade-off.

Machine learning professionals also have to deal with the bias-variance trade-off when they consider prediction problems or when choosing an algorithm to solve their classification (and regression) problems; see Rhys [95, p. 68] for a figure illustrating this trade-off and how generalization error varies with model complexity. In these situations, it is important to ask what constitutes bias and variance.

For prediction problems,

$$e = \widehat{Y} - Y$$

represents the prediction error when $\widehat{Y}$ is used to predict the target (response) variable Y. For such problems, the left-hand side of (6.1) represents average squared prediction error, and on the right-hand side, you have the variance and squared bias of the prediction error (here, bias refers to expected prediction error). As before, the trade-off between bias and variance for prediction problems is governed by (6.1).

When discussing the trade-off for classification algorithms, what does one mean when referring to bias and variance? The first thing to note is that these terms refer to properties of the *model* underlying the algorithm. Unlike the case for an estimator or a prediction, there is no precise definition of these properties for models used in machine learning. There is, however, some notional descriptions that we can give to bias and variance of a model.

Kuhn and Silge [69] refer to model bias as "the difference between the true pattern or relationships in data and the type of patterns that the model can emulate". In other words, it is the discrepancy between actual patterns in the data and their representation by a model. Models with low bias tend to be very flexible, and this quality allows them to provide good fits to various patterns in the data including those that are nonlinear and non-monotonic. Examples of such models include decision trees and k-nearest neighbors. On the other hand, models based on regression and discriminant analysis tend to have high bias because of limited flexibility and adaptability.

Model variance refers to variability of results produced by a model when supplied with the same or slightly different inputs. Linear models (such models are linear in the parameters) like linear or logistic regression usually have low variance unlike models like decision trees, k-nearest neighbors, or neural networks. The latter group of models is prone to overfitting.

When you consider the examples cited above, you see that models with low (high) bias tend to have high (low) variance. This is the bias-variance trade-off that you should take note of when deciding on choice of classifier for your problem. It helps to keep in mind that there are steps you can take to minimize the effects of the abovementioned model properties. For example, you can identify overfitting and reduce its risk by using resampling techniques [6, 95]. We take up such techniques in the next section.

6.2 Resampling Techniques for CPA

So far, we have made use of a test dataset to obtain an unbiased performance analysis of a trained classifier. As noted earlier, the resubstitution approach that uses fitted values (i.e., predictions you get when training and testing use the same dataset) to evaluate the underlying model is not a practical alternative since they tend to be optimistically biased. This is because it is not uncommon for performance measures from training data to indicate good or excellent performance, but the classifier underperforms when it is applied to unseen data like what you have in a test dataset.

However, in practice, there is often a need for unbiased evaluations of a trained classifier or its fitted model even *before* you resort to the use of a test set for this purpose; this also applies to situations when you have several competing classifiers to evaluate. Resampling techniques have an important role to play in meeting this need. Another important area where these techniques are useful is in hyperparameter tuning, i.e., the process of deciding what optimal settings to have for the classification algorithm under consideration (e.g., number of nodes for a decision tree).

In this section, we focus on the application of two resampling techniques in classification problems. We begin by showing how bootstrap resamples may be used to evaluate performance of the logit model for the Titanic classification problem. Following this, we illustrate the use of 5-fold cross-validation to tune a random forest classifier for the same problem. Finally, a brief discussion will be given that covers other resampling techniques like repeated cross-validation, Monte-Carlo cross-validation, and validation set resampling.

6.2.1 Bootstrap Resampling for Model Evaluation

The bootstrap was introduced in a seminal paper by Efron [28]. Key initial applications of this resampling scheme covered bias and standard error estimation for point estimators, and interval estimation of parameters; see Rizzo [96] for some discussion of these applications. In the last chapter, we noted that a random forest classifier relies on bootstrap resampling during training to obtain the constituent decision trees. Its use in the evaluation of classification algorithms was discussed in Cichosz [16], for example.

Given a dataset containing n cases (e.g., individuals or objects), a bootstrap resample is a related dataset of size n that is obtained by random sampling *with* replacement from the given dataset. Note that [16]

$$P(A) = 1 - \left(1 - \frac{1}{n}\right)^n, \tag{6.2}$$

where A is the event that a case is selected at least once for a bootstrap resample. For sufficiently large n, this probability is approximately 0.632. Thus, as noted in Efron and Tibshirani [30], bootstrap samples are supported by approximately $0.632n$ of the original data points.

```
# Five Bootstrap Resamples

boot_smpl
## # A tibble: 10 x 5
##      B1    B2    B3    B4    B5
##      <chr> <chr> <chr> <chr> <chr>
##  1 C     G     J     E     C
##  2 H     J     I     A     J
##  3 C     A     F     F     F
##  4 D     I     B     J     B
##  5 C     E     I     D     F
##  6 I     F     D     C     F
##  7 G     H     G     E     H
##  8 B     G     E     E     J
##  9 A     B     I     D     B
## 10 G     F     A     D     D
```

For illustration, consider the five bootstrap resamples the we just listed. The resamples were obtained by random sampling with replacement from 10 cases that were labeled using the first ten letters of the alphabet. The features and target variables for the cases are not listed since they are not relevant to the present discussion.

As shown by the column labeled **B1**, the case labeled "C" is the first one selected from the original 10 cases for the first bootstrap resample, and it appears a total of three times in the resample since sampling was done with replacement. Furthermore, because of random sampling, you get a different collection of cases in subsequent resamples. What you should note here is that the all resamples have the same size (i.e., 10) and the fact that some cases appear more than once in each resample.

What about cases not in each bootstrap resample? A case has 0.328 chance of not being included in a bootstrap resample (assuming n is sufficiently large). For classification problems, such out-of-bag cases play a useful role in model evaluations and hyperparameter tuning. For example, when evaluating a classifier, each of b bootstrap resamples plays the role of the analysis (i.e., training) set and the corresponding out-of-bag cases constitute the assessment (i.e., test) set.

```
# Out-of-Bag Cases

## $B1
## [1] "E" "F" "J"
##
## $B2
## [1] "C" "D"
##
```

The above listing shows the out-of-bag cases associated with the first two bootstrap resamples given earlier. Because of random sampling with replacement, you see a difference in number of out-of-bag cases associated with B1 and B2.

As an application, consider the use of bootstrap resampling to estimate the *Accuracy* performance parameter (as defined by the probability in Table 2.3) for the logit model classifier that is under consideration for the Titanic survival classification problem. Estimation will be based 100 bootstrap resamples from the `Titanic_train` dataset that was obtained in Section 2.2.2.

As shown in the next code segment, we begin by defining a function to compute *accuracy* for the classifier when given a training and a test dataset. Next, we use the `bootstraps()` function from the **rsample** package to obtain the splits for the resamples, and the required analysis and assessment datasets are extracted using two functions with the same name. For each iteration in the loop, this extraction is done and the *accuracy* estimate is obtained using the `acc_fn()` function.

```
# Bootstrap Point Estimation of Accuracy

# Function to Compute Accuracy Estimate for LM Classifier
acc_fn <- function(train_tb, test_tb) {
  prob_Yes <-
    glm(Survived ~ ., data = train_tb, family = binomial) %>%
    predict(newdata = test_tb) %>% plogis()
  pred_class <- ifelse(prob_Yes > 0.5, "Yes", "No") %>%
    as.factor()
  Survived = test_tb %$% Survived

  accuracy_vec(Survived, pred_class)
}

# Load the Data and Create Titanic_train

# library(rsample)
set.seed(19322)
load("Titanic_tb.rda") # Titanic_tb from Ch 2
Titanic_train <- Titanic_tb %>% initial_split() %>% training()

# Obtain List of Bootstrap Accuracy Estimates
b <- 100 # number of bootstrap resamples
set.seed(121123)
Titanic_boot_splits <-
  Titanic_train %>% bootstraps(times = b) %>% pluck("splits")
acc_boot <- list() # initialize list of bootstrap estimates
for (i in 1:b) {
  analysis_set <- Titanic_boot_splits[[i]] %>% analysis()
  assessment_set <- Titanic_boot_splits[[i]] %>% assessment()
  acc_boot[[i]] <- acc_fn(analysis_set, assessment_set)
}

# Plain Bootstrap Estimate of Accuracy
acc.plain_boot <- acc_boot %>% unlist() %>% mean()
acc.plain_boot
## [1] 0.793

# .632 Bootstrap Estimate of Accuracy
0.632*acc.plain_boot + 0.368*acc_fn(Titanic_train,Titanic_train)
## [1] 0.796
```

Given the list of *Accuracy* estimates from the bootstrap resamples, you can use a simple or weighted average (averaging aims to reduce variance) to obtain the required bootstrap estimate of *Accuracy*. The weighted version is actually an adaptation of the corresponding estimator for misclassification error that was originally proposed by Efron [29]. Evidently, the basic idea is also applicable to the problem of estimating other performance parameters. An outline of how to do it may be found in Cichosz [16], for example (an **R** function for .632 bootstrap estimation of the classfication error parameter may also be found in his article).

The first estimate of 0.793 is the simple average of *Accuracy* estimates from the bootstrap resamples. This is the result you get when you apply the plain bootstrap estimator to solve the problem. This estimator has a downward bias since each resampling estimate is effectively based on only $0.632n$ cases in the training dataset.

You can offset the abovementioned bias somewhat by averaging the estimator with the upwardly biased estimator that makes use of fitted values. The weight for the plain bootstrap estimator in the average is determined by the limiting value of the right-hand side of (6.2). The second estimate of 0.796 is the result you get when you apply this weighted averaging, i.e., it is what you get if you apply the formula given in Kohavi [67] for the .632 bootstrap estimator of *Accuracy*.

For comparison, recall from the last chapter that the *accuracy* estimate was 0.798 for the LM classifier trained on `Titanic_train` and evaluated using the `Titanic_test` (the two datasets were obtained by splitting `Titanic_tb`).

A useful alternative inferential result to consider is a confidence interval for *Accuracy* (recall, one application of such intervals is to facilitate assessment of statistical significance of differences in *Accuracy* estimates for competing classifiers). In the next code segment, we obtain two 95% bootstrap confidence intervals for this performance parameter. Discussion of these bootstrap intervals may be found in Rizzo [96], for example.

```
# Bootstrap Interval Estimation of Accuracy

# 95% Basic Percentile Interval
bpi <- acc_boot %>% unlist() %>% quantile(c(0.025,0.975))
bpi
##   2.5% 97.5%
## 0.753 0.836

# 95% Basic Bootstrap Interval
2*acc_fn(Titanic_train, Titanic_train) - rev(bpi)
## 97.5%  2.5%
## 0.769 0.852
```

6.2.2   Cross-Validation for Hyperparameter Tuning

Training a classifier in practice often requires you to tune one or more hyperparameters to achieve optimal performance. Tuning is the process that you

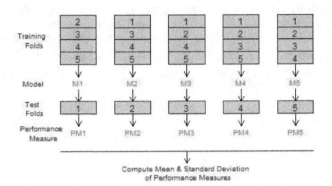

FIGURE 6.1
Illustrating 5-Fold Cross-Validation

use to find the "best" setting for the hyperparameters of a classification algorithm. This is often accomplished through the use of a suitable resampling technique.

When you use cross-validation for hyperparameter tuning, the process begins by splitting all available data for classifier construction into training and test sets. For k-fold cross-validation, you next split the training set into k disjoint subsets of roughly equal size; these subsets are referred to as folds. You can associate a resample with each of these k folds. The i-th fold serves as the assessment set for the i-th resample, and the remaining $k-1$ folds serves as analysis set. A good illustration of the overall process that is involved may be found in Figure 3.5 of Kuhn and Johnson [68, p. 48]; B in their figure corresponds to our k.

For further illustration, see Figure 6.1 which shows the resamples for 5-fold cross-validation; the figure was adapted from Provost and Fawcett [90, p. 128]. In each of the resamples, the training folds make up the analysis set and the test fold constitutes the assessment set. When used for model evaluation or hyperparameter tuning, you train each of the five models using the same classification algorithm and corresponding analysis set, and evaluate each trained model using the corresponding test fold and a performance measure like *accuracy* or *AUC*; see Cichosz [16, p. 150] for an example on the use of 10-fold cross-validation to evaluate a decision tree classifier. The discussion below on the use of 5-fold cross-validation to tune the hyperparameters of a random forest classifier is another example.

In general, the number of hyperparameters you need to tune depends on the software implementation of a classification algorithm. For example, when you use the `rand_forest()` function in the **parsnip** package to train a random forest classifier, the relevant hyperparameters are given by:

- `trees`, the number of trees in the ensemble (default = 500),

- `min_n`, the minimum number of cases in a node for further splitting,

- `mtry`, the number of randomly selected predictors at each split for a constituent decision tree.

For classification, the default for `min_n` and `mtry` are 10 and $\lfloor\sqrt{p}\rfloor$, respectively, where p is the number of predictors. A discussion of the roles played by the above hyperparameters in the training of an RF classifier may be found in Boehmke and Greenwell [6, p. 206], for example. Here, we demonstrate the use of 5-fold cross-validation (in practice, using 10 folds is quite common) to tune these hyperparameters for the problem in Section 2.2.

We begin by splitting `Titanic_df` (instead of `Titanic_tb`, as was done in the second chapter) to create the training and test datasets. This is done so that we can demonstrate how to handle the missing values in `Titanic_df` through the use of a preprocessing recipe. When you take this preferred approach, the required packages are loaded when you load the **tidymodels** meta-package.

```
# Recreate Training and Test Datasets for the DT Classifier

library(tidymodels) # loads several core packages
set.seed(19322)
Titanic_split <- Titanic_df %>% initial_split()
Titanic_train <- Titanic_split %>% training()
Titanic_test <- Titanic_split %>% testing()

# Model Specification
rf_mod_spec <-
    rand_forest(mtry = tune(), min_n = tune(), trees = tune()) %>%
    set_mode("classification") %>%
    set_engine("randomForest")

# Preprocessing Recipe
rf_rec <-
    recipe(Survived ~ ., data = Titanic_train) %>%
    step_impute_median(Age) # impute missing values

# Workflow
rf_wf <-
    workflow() %>%
    add_model(rf_mod_spec) %>%
    add_recipe(rf_rec)
```

In the model specification step, the hyperparameters to be tuned are identified by using the `tune()` placeholder; also specified is the type of problem to be solved and the **R** implementation of the algorithm to use. The preprocessing recipe specifies the use of the median of non-missing `Age` values to impute the missing ones. The objects representing the two specifications are then incorporated into a `"workflow"` object that will be used when tuning the hyperparameters.

The setup phase of the tuning process starts with the required preprocessing to obtain `Train_data` (here, this is `Titanic_train` with missing values imputed). This is followed by creating training/test folds for the 5-fold cross-validation and a grid of 75 (= 3 × 5 × 5) hyperparameter combinations.

```
# Tune Hyperparameters for a RF Classifier

# Obtain Processed Training Data
Train_data <- rf_rec %>% prep() %>% juice()

# Create Training & Test Folds
set.seed(25322)
rf_folds <- vfold_cv(Train_data, v = 5)

# Create Grid of Hyperparameter Combinations
set.seed(18422)
rf_grid <-
  grid_regular(mtry(range = c(2, 4)), min_n(range = c(2,20)),
    trees(range = c(500, 2000)), levels = c(3,5,5))
```

To complete the tuning process, a grid-search is performed to find the combination of hyperparameters that yields the "best" value for a given metric (e.g., *accuracy*). The basic algorithm to solve this optimization problem comprises two loops as shown below.

```
# Grid-Search Algorithm for RF Hyperparameter Tuning
# with 5-Fold Cross-Validation

for each combination of hyperparameters in rf_grid do
|  for each resample in rf_folds do
|  |  Use the analysis set to obtain a RF classifier
|  |  Use the classifier and assessment set to obtain predictions
|  |  Compute the performance metric
|  end
|  Average the computed metrics and obtain the standard error
end
Sort hyperparameter combinations by the computed averages
Return the combination with "best" value for the metric
```

You can, of course, write your own code to implement the algorithm that was outlined above. To do this, you need to flesh out details of the algorithm and code it in the programming language of your choice.[4] You have another option, i.e., you can use the `tune_grid()` function from the **tune** package to obtain the required results as shown below.

---

[4]For some idea of what is involved when using **R**, see the *k*-fold cross-validation code in Cichosz [16, p. 150]. The given code is quite instructive even though it is for model evaluation rather than hyperparameter tuning.

```
# Tune Hyperparameters for a RF Classifier (cont'd)

# Obtain Test Fold Predictions
rf_rs <- rf_wf %>%
  tune_grid(resamples = rf_folds, grid = rf_grid,
    control = control_grid(save_pred = TRUE))
rf_rs
## # Tuning results
## # 5-fold cross-validation
## # A tibble: 5 x 5
##    splits             id    .metrics .notes   .predicti~1
##    <list>             <chr> <list>   <list>   <list>
## 1 <split [534/134]> Fold1 <tibble> <tibble> <tibble>
## 2 <split [534/134]> Fold2 <tibble> <tibble> <tibble>
## 3 <split [534/134]> Fold3 <tibble> <tibble> <tibble>
## 4 <split [535/133]> Fold4 <tibble> <tibble> <tibble>
## 5 <split [535/133]> Fold5 <tibble> <tibble> <tibble>
## # ... with abbreviated variable name 1: .predictions
```

Results from tuning are collected into the nested tibble `rf_rs`. The listing of this tibble shows that the first three resamples contain 534 cases in each set of training folds and 134 cases in each test fold. The corresponding numbers for the last two resamples are 535 and 133, respectively. The `.metrics` column is a list-column of 5 tibbles, each of which contains *accuracy* and *AUC* values for the 75 hyperparameter combinations. Corresponding test fold predictions are contained in the list column labeled `.predictions`. Use the `collect_metrics()` or `collect_predictions()` functions if you want the details for the various hyperparameter combinations. The second function will be used later to extract the test fold predictions for the "best" model; see definition of the `best_rf_rs_pv` tibble that is given later.

You now have sufficient information to determine the "best" combination of hyperparameters to use for the required RF classifier. The combination depends on the metric used to make the selection. In general, you get different combinations with different metrics. As noted earlier, you might even get different combinations with the same maximal value for a given performance metric.

```
# Tune Hyperparameters for a RF Classifier (cont'd)

# "Best" Hyperparameter Combination
rf_rs %>% select_best(metric = "accuracy")
## # A tibble: 1 x 4
##    mtry trees min_n .config
##    <int> <int> <int> <chr>
## 1    3   875     6 Preprocessor1_Model20
```

As shown in the preceding code segment, the hyperparameter combination for the "best" model based on the *accuracy* metric is given by

$$\texttt{mtry} = 3, \texttt{trees} = 875 \text{ and } \texttt{min\_n} = 6.$$

This combination of the three hyperparameters is for the model designated as `Preprocessor1_Model20`. The mean *accuracy* for this model is 0.820 and the corresponding standard error is 0.0111. These values are summaries from estimates of *accuracy* that were obtained from predictions that resulted when this model was applied to data in the five test folds. We provide details in the discussion following the next code segment.

```
# Tune Hyperparameters for a RF Classifier (cont'd)

# Mean and Std Error of accuracy by Hyperparameter Combination
rf_rs %>% show_best("accuracy") %>%
  select(mtry, trees, min_n, .config, mean, std_err)
## # A tibble: 5 x 6
##     mtry trees min_n .config                    mean std_err
##    <int> <int> <int> <chr>                     <dbl>   <dbl>
## 1      3   875     6 Preprocessor1_Model20     0.820  0.0111
## 2      3  1250     6 Preprocessor1_Model35     0.820  0.00665
## 3      4  1250     6 Preprocessor1_Model36     0.819  0.00800
## 4      3   500     6 Preprocessor1_Model105    0.817  0.00860
## 5      3  2000     6 Preprocessor1_Model165    0.817  0.00684
```

Note from the above results that there is a second combination of hyperparameters that yields the same mean *accuracy* value. To resolve the non-uniqueness problem, you need to consider some other criteria (e.g., preference for smaller number of trees).

For illustration, we show how to obtain the mean and standard error of average *accuracy* for the first model in the tibble listed above. To do this, you first need to use the `collect_predictions()` function to extract the predictions from the `rf_rs` tibble that was obtained earlier, and then follow up by using the `filter()` function to extract the resampling results for the model.

```
# Mean accuracy and Standard Error for the "Best" Model

# Extract Test Fold Predictions for the "Best" Model
best_rf_rs_pv <-
  rf_rs %>%
  collect_predictions() %>%
  filter(.config == "Preprocessor1_Model20") %>%
  select(id, mtry, trees, min_n, .pred_Yes, .pred_class,
    Survived) %>%
```

```
 mutate_if(is.factor, fct_rev)

best_rf_rs_pv %>% print(n = 3)
## # A tibble: 668 x 7
##   id     mtry trees min_n .pred_Yes .pred_class Survi~1
##   <chr> <int> <int> <int>     <dbl> <fct>       <fct>
## 1 Fold1     3   875     6     0     No          No
## 2 Fold1     3   875     6     0     No          No
## 3 Fold1     3   875     6     0.999 Yes         Yes
## # ... with 665 more rows, and abbreviated variable
```

Given the resampling results for `Preprocessor1_Model20` that are contained in the best_rf_rs_pv tibble, you can obtain estimates of *accuracy* from predicted and actual classes in each test fold, and then obtain the required summaries from the resulting five *accuracy* estimates.

```
# Mean accuracy and Standard Error for the "Best" Model (cont'd)

# Obtain Test Fold Accuracy Measures & Summary Statistics
best_rf_rs_pv %>%
  group_by(id) %>%
  summarize(acc = accuracy_vec(Survived, .pred_class)) %T>%
  print() %>%
  summarize(Mean = mean(acc), StdErr = sd(acc)/sqrt(5))
## # A tibble: 5 x 2
##   id      acc
##   <chr> <dbl>
## 1 Fold1 0.813
## 2 Fold2 0.828
## 3 Fold3 0.813
## 4 Fold4 0.857
## 5 Fold5 0.789
## # A tibble: 1 x 2
##    Mean StdErr
##   <dbl>  <dbl>
## 1 0.820 0.0111
```

Information in the best_rf_rs_pv tibble may also be used to obtain the test fold ROC curves in Figure 6.2 and the corresponding *AUC* values (including the summary statistics). The displayed curves show quite a bit of variation. This variation is reflected in the corresponding *AUC*s. As shown below, the

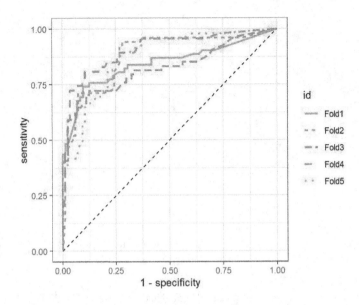

FIGURE 6.2
ROC Curves from the Test Folds for the "Best" Model

values range from 0.824 to 0.906. The average of the five AUCs is 0.865 with a standard error of 0.0143. We leave it to Exercise 3 for the reader to verify these results.

```
# Test Fold AUCs for the "Best" Model with Summary Statistics

## # A tibble: 5 x 2
##    id      auc
##    <chr> <dbl>
## 1 Fold1 0.844
## 2 Fold2 0.879
## 3 Fold3 0.824
## 4 Fold4 0.906
## 5 Fold5 0.871
## # A tibble: 1 x 2
##    Mean StdErr
##   <dbl>  <dbl>
## 1 0.865 0.0143
```

The "best" hyperparameter combination may be used to obtain a final fit for the RF classifier. One way to do this is to proceed as shown below.[5] You can analyze performance of the resulting trained classifier as demonstrated in earlier chapters by using information in `Test_data`. This is also left as an exercise for the reader.

```
# RF Classifier with "Best" Hyperparameter Combination

# Fit the RF Classifier Using Tuned Hyperparameters
rf_fit <-
  rand_forest(mtry = 3, trees = 875, min_n = 6) %>%
  set_mode("classification") %>%
  set_engine("randomForest") %>%
  fit(Survived ~ ., data = Train_data)

# Processed Test Data
Test_data  <- rf_rec %>% prep() %>% bake(Titanic_test)
```

6.2.3 Other Resampling Schemes

There are several other ways you can use resampling to evaluate performance of your model. In this section, we give a brief outline of some of the more useful alternatives; see James et al. [62] or Kuhn and Silge [69] for more information. The **rsample** package contains the required functions to implement these resampling techniques.

When k-fold cross-validation yields resampling estimates of performance measures with too much variability, you can use repeated cross-validation to reduce the standard error of the average value you obtain from the resampling estimates. The estimated standard error of the average when you use k-fold cross-validation is $\hat{\sigma}/\sqrt{k}$, where $\hat{\sigma}$ is the estimated standard deviation of the k resampling performance estimates. When you repeat this cross-validation m times, the standard error of the average is reduced to $\hat{\sigma}/\sqrt{m \times k}$. However, based on experimental results and reasons given in their article, Vanwinckelen and Blockeel [110] noted that "repeated cross-validation should not be assumed to give much more precise estimates of a model's predictive accuracy". This does not mean that the resampling scheme is not useful; it means that you need to address some of the concerns they have for making this claim.

For the earlier example on random forest hyperparameter tuning, we can consider 10 repetitions of the 5-fold cross-validation. This can be done by adding `repeats = 10` as part of the argument in the previous call to the `vfold_cv()` function; see the earlier command to create `rf_folds`. In the

---

[5]Alternatively, you can use the `last_fit()` function to fit the final model using the training data after you update the model specification and workflow, and evaluate it with the test dataset; see Kuhn and Silge [69], for example.

presence of serious class imbalance, it is recommended that you use strati-
fied k-fold cross-validation; see Brownlee [10] for discussion of this issue. The
vfold_cv() function allows you to do this through its **strata** argument.

A viable resampling technique when you have a large dataset is the validation set approach. In this approach, you effectively partition the dataset
into three subsets for use in training, validation, and testing. The partitioning may be performed using a three-way initial split or by an initial two-way
split followed by a subsequent split of the portion not set aside for testing; see
Kuhn and Silge [69] for a couple of figures that illustrate the two possibilities.
When resampling for tuning or performance evaluation, you repeat random
splits of cases into training and validations sets, and then fit the model using
each training set and use the corresponding validation set to estimate performance measures. As an example of this approach, see the case study at
https://www.tidymodels.org/start/case-study/ on its use to tune and
train a penalized logistic regression model.

Finally, you can also consider Monte Carlo cross-validation. For this approach, you obtain a resample by taking a fraction of the original dataset by
random sampling without replacement to make up the analysis set, and the
remaining data as assessment set. You can use the mc_cv() function from the
rsample package to obtain the splits that you require for the cross-validation;
see Kuhn and Silge [69, p. 134] for an example on how to do this.

6.3   Dealing with Class Imbalance

Many classification problems that you encounter in practice are imbalanced
because of significant differences in sizes of the classes involved in the problem. Examples may be found in areas like bioinformatics, fraud detection,
information retrieval, medical diagnosis, and spam filtering. Such imbalance
complicates the classifier development process in a number of ways. They
impact not only the training process but also what you need to do when evaluating the trained classifier. During training, severe class imbalance means
that there are very few cases in the minority class, e.g., fraudulent transaction constitute only 0.173% of the creditcardfraud dataset from **kaggle**.[6]
This affects the estimation/fitting process for classifiers that are not robust in
the presence of class imbalance. When it comes to CPA, you have to choose
suitable performance measures to evaluate your classifier since certain obvious and intuitive measures like *accuracy* have questionable value when class
imbalance is serious.

The problem basically arises when there are significant differences in the
sizes of classes due to nonuniformity in the class distribution underlying the

---

[6]Source: https://www.kaggle.com/datasets/mlg-ulb/creditcardfraud

substantive problem (e.g., credit card fraud usually occurs at relatively low frequency) and the data you have for classifier training is representative of the population it came from.

How serious is the class imbalance before you take steps to account for it when you train and evaluate your classifier? To answer this question, you have to first decide whether the problem exists, and if so, proceed to judge its seriousness. The existence decision for binary classification may be based on the imbalance ratio (IR) that you can obtain by dividing the number of cases in the *negative* class by the number in the *positive* class (this usually refers to the minority class); this is also known as the class or skew ratio [40].

As a rough rule of thumb, class imbalance can be said to exist if the IR is more than 1.5. According to this rule, the `Titanic_train` dataset that we used in Section 2.2.2 is imbalanced since its IR is 1.54, but is it concerning? The answer is probably not since the difference between it and the suggested cutoff is negligible. For comparison, consider the five project datasets in the book by Brownlee [10] on imbalanced classification. The IR for these datasets ranges from 2.33 to 42. The `creditcardfraud` dataset mentioned earlier is a more extreme example since $IR = 578$ for it.

In this section, we provide some background information on class imbalance issues and a brief overview of some techniques to deal with the problem. For a more complete coverage of the available techniques, see Fernández [36], for example. Readers with background in **Python** may find the tutorials in Brownlee [10] a useful resource. His tutorial on a systematic framework for dealing with imbalanced classification problems is noteworthy.

6.3.1   CPA when Class Imbalance is Present

How does one deal with the class imbalance problem? There are several aspects to consider when answering this question. From the CPA perspective, the problem requires you to make a careful selection of measures when evaluating classifier performance. For binary classification, the selected measure depends on factors like the prediction problem under consideration (class labels versus class membership probabilities) and relative importance of the two classes. As part of his detailed framework for working through an imbalanced classification project, Brownlee [10, p. 308] provided a flowchart that encapsulates the metric selection process. It shows that, among others, measures like F_β-score (for $\beta = 0.5, 1, 2$) and AUC (from ROC and PR curves) are possible choices when you have class imbalance.[7] Notice also from his flowchart that *accuracy* is still a viable measure provided you have a class label prediction problem with equally important classes and the majority class constitutes less than $80 - 90\%$ of the dataset (it suffices to require that $IR < 9$).

There are also aspects that are relevant to classifier performance to consider although they do not have a *direct* bearing on CPA (like choice of

---

[7]Note that Brownlee [10] uses the term F_β-score rather than F_β-measure.

classification algorithm). Standard classification algorithms like those based on decision trees and multi-layer perceptrons are not robust. While they work well for balanced classification problems, their performance are negatively impacted by the presence of class imbalance. To overcome the deterioration in performance, you can consider either data or classifier level approaches to deal with the problem. For completeness, it is useful to examine these topics because understanding important issues related to the broader area of classifier performance is essential for a more comprehensive approach to CPA.

6.3.2 Data-Level Approaches

Data-level approaches to resolve the class imbalance problem involve the use of suitable sampling techniques to alter the class distribution in the *training* dataset for classification algorithms that do not work well when you have the problem. The change in distribution may be achieved through the use of over-sampling or under-sampling techniques, or a combination of the two.

With over-sampling, you can use one of several techniques to increase the size of the minority class. A simple approach is to add randomly selected cases from this class. This random over-sampling technique is prone to overfitting, but according to Batista et al. [4], the technique competes well with more complex over-sampling methods. Alternatively, you can perform the required data augmentation by interpolating neighboring cases in the minority class through the use of the Synthetic Minority Over-sampling Technique (SMOTE) technique proposed by Chawla et al. [13].

To illustrate, consider the `Titanic_train` dataset that we used to train the decision tree classifier in Section 2.2.2. This training dataset is imbalanced, although not to a great degree (the IR is about 1.5). Next, we recreate this dataset and demonstrate the use of the `smote()` function from the **performanceEstimation** package to balance it by over-sampling. The result we get is the `Titanic_Train` dataset in which the size of the minority class is increased to three times the original number.

```
# Balance Training Data for Titanic Survival Classification

# Recreate the Training Dataset
library(rsample)
set.seed(19322)
Titanic_train <-
  Titanic_tb %>% initial_split() %>% training()
Titanic_train %$% table(Survived) %T>% print() %>% prop.table()
## Survived
##  No Yes
## 405 263
## Survived
##    No   Yes
```

```
## 0.606 0.394

# Obtain a Balanced Training Dataset Using SMOTE
library(performanceEstimation)
perc_over <- 2; perc_under <- (1 + perc_over) / perc_over
Titanic_Train <-
  smote(Survived ~ ., data.frame(Titanic_train),
      perc.over = perc_over, perc.under = perc_under)
Titanic_Train %$% table(Survived) %T>% print() %>% prop.table()
## Survived
##  No Yes
## 789 789
## Survived
##  No Yes
## 0.5 0.5
```

Under-sampling, on the other hand, allows you to reduce the size of the majority class by removal (of redundant, borderline, or noisy cases) or retention (of useful cases). For example, deletion can be done by random under-sampling, i.e., randomly eliminate cases from the majority class. An obvious issue with this non-heuristic approach is the potential loss of information when you remove useful cases (i.e., those that facilitate classifier training), but there are techniques to overcome this problem (e.g., data decontamination by relabeling some cases).

Heuristics methods are also available for under-sampling like the one that makes use of pairs of cross-class nearest neighbors, the so-called Tomek links [109]. By removing cases that belong to the majority class from each cross-pair, you increase separation between the classes. Such pairs may be combined with SMOTE to create class clusters that are better defined [4]; see Figure 2 in the cited article for a good demonstration of the basic idea underlying this hybrid approach.[8] Another heuristic you can use is to apply the condensed nearest neighbor rule [58] to keep cases in the majority class that are useful together with all cases in the minority class. The resulting dataset is referred to as a consistent subset.

6.3.3   Classifier Level Approaches

Instead of modifying the training dataset to deal with class imbalance, classifier level method modify the classification algorithm. In general, such modifications depend on the algorithm in question. Determining the mechanism that causes the bias towards the majority class is an important first step to consider when you want to modify an algorithm to work in the presence of

---

[8]This approach illustrates an important point, namely, the fact that class imbalance is not the only obstacle to classifier induction, but other factors like overlapping data from different classes can also be problematic.

TABLE 6.1
Node Class Counts Before and After a
Binary Split

Node	#Positive	#Negative
Parent	n_P	n_N
Left Child	n_{LP}	n_{LN}
Right Child	n_{RP}	n_{RN}

class imbalance. For example, the critical mechanism responsible for the bias in tree-induction algorithms is the splitting criterion.

Decision trees with splits determined by measures of impurity based on Gini index or entropy (as used by the **CART** or **C4.5** algorithm, respectively) perform poorly when you have class imbalance. The issue is more problematic with use of Gini index as demonstrated by the isometrics of Gini-split in Figure 6(b) of Flach [40]. You can obtain skew-insensitive performance by using an alternative splitting criterion. For binary classification, a good alternative is to split based on Hellinger distance. This criterion may be expressed as

$$d_H = \sqrt{\left(\sqrt{\frac{n_{LP}}{n_P}} - \sqrt{\frac{n_{LN}}{n_N}}\right)^2 + \left(\sqrt{\frac{n_{RP}}{n_P}} - \sqrt{\frac{n_{RN}}{n_N}}\right)^2}, \quad (6.3)$$

where the counts on the right-hand side are given in Table 6.1. The above formula may be expressed as [17]

$$d_H(tpr, fpr) = \sqrt{\left(\sqrt{tpr} - \sqrt{fpr}\right)^2 + \left(\sqrt{1-tpr} - \sqrt{1-fpr}\right)^2}.$$

For this to make sense, you need to adopt another interpretation of the counts in the second and third rows of Table 6.1, i.e., regard them as counts from a confusion matrix with key count vector $(n_{LP}, n_{RP}, n_{LN}, n_{RN})$. Hence, the definitions of tpr and fpr in $d_H(tpr, fpr)$ follow from (1.11) and (1.14) in Chapter 1 when you interpret this vector according to what is given by (1.9).[9] One advantage of the alternative formula is that it allows you to obtain isometric plots for Hellinger distance as shown in Figure 1 of Cieslak et al. [18, p. 141]. The given figure demonstrates the robustness of this distance measure in the presence of skew.

For tree induction, d_H is a useful splitting criterion since it is a good measure of the ability of a feature to separate the classes. An algorithm to construct a Hellinger distance decision tree (HDDT) that makes use of this criterion may be found in Cieslak and Chawla [17]. Based on experiments they carried out, Cieslak et al. [18] concluded that HDDTs are robust and

---

[9]Here, rather than think of tpr and fpr as classifier performance measures, view them simply as relative frequencies derived from the confusion matrix representation of class distributions in the child nodes that result from a decision tree split.

applicable to a wide variety of datasets, thus reinforcing the skew-insensitive property of HDDTs that was highlighted in [17]. For an application of HDDT, see Pozzolo et al. [89].

The above discussion provides a glimpse of what is involved when you attempt a classifier level approach to modify a particular learner for class imbalance problems. Modifications are also available for other classification algorithms like those based on support vector machines, nearest neighbor, and naive Bayes; see Fernández et al. [36] for a survey of these techniques.

Finally, note that you may also regard cost-sensitive learning as a classifier level approach to solving problems with class imbalance. This is a significant current research area, and a brief overview of this topic is given in the next section.

6.3.4 Cost-Sensitive Learning

Unequal costs for misclassification errors is an important issue when you have class imbalance, and you need to account for the difference in costs when solving such problems. For example, false negatives in binary problems are often more serious than false positives (e.g., misclassifying a fraudulent case is more serious than wrongly classifying a non-fraudulent one). By making misclassification of minority cases more costly, cost-sensitive learning provides another approach to deal with the class imbalance problem. A key aspect of this approach is the use of a cost matrix.

For binary classification, the cost matrix may be expressed as

$$C = \begin{bmatrix} c_{11} & c_{12} \\ c_{21} & c_{22} \end{bmatrix} \qquad (6.4)$$

where c_{ij} is the cost incurred when a class j case is classified as class i (here, class 1 refers to the *positive* class and, for imbalanced problems, this coincides with the minority class). The c_{ij}'s represent values of a suitable *measure* of cost which, in general, need not be in monetary terms. The off-diagonal entries in C are particularly important because c_{12} and c_{21} refer to the cost of a false positive and the cost of a false negative, respectively. Since incorrect labeling of a case should cost more, a requirement for c_{ij}'s to satisfy is

$$c_{21} > c_{11} \text{ and } c_{12} > c_{22}.$$

These are the "reasonableness" conditions (in different notation) given in Elkan [32].

In general, specifying the entries in (6.4) is difficult. One heuristic that is useful for some problems makes use of the following specification:

$$c_{11} = 0, c_{12} = k, c_{21} = k \times IR \text{ and } c_{22} = 0.$$

This IR-based heuristic accounts for more costly false negatives by making cost of such errors proportional to the imbalance ratio.

You can set $k = 1$ in the abovementioned heuristic without loss of generality because re-scaling (and shifting) entries of a cost matrix does not change optimal decisions based on it [32]. Thus, you can also simplify cost matrix $\boldsymbol{C}$ by using c_{22} and $(c_{12} - c_{22})$ to, respectively, shift and re-scale entries in $\boldsymbol{C}$ to obtain

$$\boldsymbol{C} = \begin{bmatrix} \frac{c_{11}-c_{22}}{c_{12}-c_{22}} & 1 \\ \frac{c_{21}-c_{22}}{c_{12}-c_{22}} & 0 \end{bmatrix} = \begin{bmatrix} \tilde{c}_{11} & 1 \\ \tilde{c}_{21} & 0 \end{bmatrix}.$$

To illustrate the use of $\boldsymbol{C}$, consider binary probabilistic classifiers which employ rules of the form:

Assign a case to the *positive* class if $P(Y = 1 \mid \boldsymbol{x}) > t$,

where, as in in Section 1.1.2 of Chapter 1, $\boldsymbol{x}$ and Y are the feature vector and corresponding target variable of a case. You can take a cost-sensitive approach to determine the threshold t as shown below (the default $t = 0.5$ is not optimal when you have class imbalance).

Based on expected costs derived from $\boldsymbol{C}$, the optimal prediction is class 1 for a case if and only if

$$p\tilde{c}_{11} + (1 - p) \leq p\tilde{c}_{21},$$

where p is equal to $P(Y = 1 \mid \boldsymbol{x})$. The left-hand side of the above inequality is the expected cost from the prediction that a case is in class 1, and the right-hand side is the corresponding expectation for class 2. On re-arranging terms, we have

$$p \geq \frac{1}{\tilde{c}_{21} - \tilde{c}_{11} + 1} = \frac{c_{12} - c_{22}}{c_{21} - c_{11} + c_{12} - c_{22}} = p^*.$$

Thus, p^* is the cost-sensitive threshold for a probabilistic classifier that minimizes expected misclassification cost. Elkan [32] derived p^* using the cost matrix similar to (6.4) instead of the simpler $\boldsymbol{C}$. The fact that we arrive at the same p^* demonstrates the earlier remark regarding re-scaling and shifting entries in a cost matrix.

When the cost matrix is determined by the IR-based heuristic with $k = 1$,

$$p^* = \frac{1}{IR + 1} = \frac{n_1}{n_1 + n_2}.$$

In this case, the optimal threshold is simply the prevalence of the minority class. In particular, when classes are balanced, $p^* = 0.5$.

Thus, when you use the standard LM classifier that makes *positive* class assignment based on (1.4) in Chapter 1, you can set $c = p^*$ to obtain a cost-sensitive classification rule. However, a more common approach in practice to obtain such a rule is to estimate the model parameters by minimizing a weighted cost function. This function is given by the weighted negative log-likelihood function

$$\ell(\boldsymbol{\beta}) = -w_1 \sum_{i=1}^{n} z_i \ln p_i - w_2 \sum_{i=1}^{n} (1 - z_i) \ln(1 - p_i), \qquad (6.5)$$

where β is a vector of LM model parameters, w_1 and w_2 are the weights,

$$z_i = 2 - y_i \text{ and } p_i = \frac{\exp(\eta(\boldsymbol{x}_i))}{1 + \exp(\eta(\boldsymbol{x}_i))},$$

for $i = 1, \ldots, n$. Note that the y_i's are the observed values of the target variable (recall $y_i \in \{1,2\}$ which explains why we transform them to z_i), and $\eta(\cdot)$ is the function that defines the linear predictor; see (1.5) or (4.1) for examples of this function. The first (second) term on the right-hand side of (6.5) is the contribution to cost by cases in the minority (majority) class.

How does one determine the weights? Reliance on domain experts is the obvious approach to take. Or, you can also treat the issue as a hyperparameter tuning problem. A simpler and sometimes useful alternative is to use the formula

$$w_i = \frac{n/2}{n_i}, \quad i = 1, 2,$$

where n_i is the size of class i and $n = n_1 + n_2$. This formula follows from the "best practice heuristic" in Brownlee [10, p. 195]. It is equivalent to setting $w_1 = IB \times w_2$. Thus, if you know the IB for your training data, one option is to set $w_1 = IB$ and $w_2 = 1$ in (6.5).

For further discussion on cost-sensitive learning, see Fernández et al. [36]. Some tutorials on this topic may be found in Brownlee [10].

6.4 Summary and Conclusions

At an elementary level, classifier performance analysis seems quite a simple thing to do. Apparently, to perform the analysis, it suffices to calculate some relevant intuitive measures like fraction of cases that were correctly classified in an evaluation dataset (i.e., *accuracy*), fraction of *positive* cases that were correctly classified (i.e., *sensitivity*), and fraction of correct *positive* classifications (i.e., *precision*). Then, choose one or more of these measures for the performance analysis you plan to do. If you want to do more, you can also take into account the ROC curve and the area under it (i.e., *AUC*).

Unfortunately, the above selection of metrics is somewhat limited, and not all intuitive and apparently reasonable measures work as expected in practice. Practical issues like class imbalance can diminish the utility of a performance measure (e.g., *accuracy*). There are also subtle issues like the ability of a measure to coherently quantify performance of a classifier. The widely used *AUC* ranking measure is a case in point. Its incoherency has been highlighted by Hand [50]. An issue like this is not what some analysts and researchers think about, given the interest they still have with this particular metric.

The research by Chicco and Jurman [15] led them to choose *mcc* over *AUC* as the standard metric for binary CPA. The same comparison was made by

Halimu et al. [45], but they concluded that AUC is a better measure. Huang and Ling [61] compared AUC and *accuracy* in their study, and argued that optimizing AUC is a better option in practice. Unfortunately, these studies have one problem with them. In terms of the taxonomy proposed by Ferri et al. [37], these studies involve comparisons between threshold and ranking measures. Hence, the measures being compared quantify different aspects of classifier performance and, as noted by Hand [52], empirical studies engaged in such comparisons have limited value.

In an earlier study, Chicco and Jurman [14] compared three commonly used threshold measures and concluded that *mcc* is more reliable compared to F_1-measure and *accuracy*. The shortcomings of *accuracy* have been known for some time. It is not a useful performance measure when class imbalance is serious, and when the cost of misclassifications matter. Furthermore, its use in comparative performance analysis of competing classifiers in not recommended as noted by Provost et al. [91]; they preferred the use of ROC analysis instead. In their study involving *mcc* and *kappa*, Delgado and Tibau [22] highlighted the incoherence of *kappa* and argued against the use of this measure. Thus, of the three overall threshold measures that we considered in the second chapter, the available evidence seems to favor *mcc*.

There are also issues that arise with class-specific measures. For some problems, the goal is to minimize both the false discovery of *positive* cases and occurrence of false negatives. In this situation, the relevant measures to consider are *precision* and *recall*, and possibly F_1-measure, their harmonic mean. Powers [88] noted some biases associated with these measures, e.g., failure to provide information about how well *negative* cases are handled by a classifier. Their use to evaluate classifiers for information retrieval systems can be problematic when retrieval results are weakly ordered as noted by Raghavan et al. [93]; see their article for probabilistic measures designed to deal with the problem. Hand and Christen [46] highlighted a conceptual weakness of F_1-measure when used to evaluate record linkage algorithms and suggested alternative measures to overcome the problem.

If minimizing both the occurrence of false negatives and occurrence of false positives is an important goal, then *sensitivity* and *specificity* are the measures to consider, including composite measures derived from them like J-index or *discriminant power*. If you want more than a pair of estimates of the first two measures, then you can consider the ROC curve; this curve is determined (directly or indirectly) by these measures. The area under the ROC curve provides a useful descriptive summary for a single classifier, but issues arise if you try to compare classifiers using AUC. In the latter situation, the H-measure proposed by Hand [50] provides a solution to the problem.

There are alternatives to the ROC curves. One important example is the Precision-Recall (PR) curve. According to Davis and Goadrich [19], these curves provide a more informative view of a classifier's performance when you have significant class imbalance. This assessment is supported by Saito and Rehmsmeier [99].

So far in our discussion, we focused attention on CPA for binary classifiers. There are added complications when you consider multiclass CPA. While some performance measures apply to both types of analysis (e.g., *accuracy*), others require adoption of an OvR perspective (this yields multiclass class-specific measures and *AUC*). ROC hypersurfaces and volumes under them have limited utility when the number of classes is too large.

One conclusion you can draw from the remarks in this section is the observation that CPA is a challenging task.[10] There are a variety of issues and trade-offs to keep in mind when making the analysis. Commonly used and intuitive measures may fail to work for certain problems while some of the other measures suffer from incoherence.

The performance measures and curves covered in this book represent the most common metrics used in CPA. For obvious reasons, there is a limit to how much can be covered in an introductory text. There are other metrics in the published literature that you can consider using for your problem, e.g., confusion entropy [21, 64, 114] for multiclass problems. Hopefully, with the knowhow acquired from reading this book and that from the cited references, you can confidently explore some of these alternatives to see if they can do a better job at evaluating classifier performance. The underlying message is this: as a subject, CPA is still evolving and there is much to learn, and **R** is an excellent aid for learning and applying what you need to know.

6.5 Exercises

1. Let x_{min}, x_{max}, $\bar{x}$, and s denote the minimum, maximum, mean, and standard deviation obtained from a numerical data sample that is represented by $x_1, \ldots, x_n$. To normalize the x_i's, you perform the following transformation(s)

$$x_i \mapsto \frac{x_i - x_{min}}{x_{max} - x_{min}}, \quad i = 1, \ldots, n,$$

and to standardize the x_i's, you perform

$$x_i \mapsto \frac{x_i - \bar{x}}{s}, \quad i = 1, \ldots, n.$$

During our discussion on feature engineering, we showed how to use the **recipes** package to normalize and standardized features. Show how to perform these transformations without using the stated functionality. Use the **mpg_tb** tibble that was created in text example.

---

[10]This holds even though we made no mention of some of the other important topics in this chapter like resampling techniques for CPA.

2. For this exercise, consider the LM classifier in the Chapter 4 for the Titanic survival classification problem in Section 2.2.

   (a) Interpret the probability given on the right-hand side of (3.2) for the classification problem under consideration.

   (b) Obtain the plain and .632 bootstrap estimate of the probability mentioned in part (a).

   (c) Obtain a basic percentile interval and a basic bootstrap interval for the probability. Use a 90% confidence level.

3. From the example on hyperparameter tuning for a random forest classifier by 5-fold cross-validation, we noted that the model identified by `.config == "Preprocessor1_Model20"` is the "best" one for the problem in Section 2.2 according to the *accuracy* measure.

   (a) Show how to obtain Figure 6.2 and the corresponding *AUC*s. Also, verify the given summary statistics for the *AUC* values.

   (b) Obtain a list of the five test fold confusion matrices for this model.

   (c) Obtain the five test fold values for Cohen's *kappa*, hence, the mean and standard error from these values.

4. Consider the same RF hyperparameter tuning example as used for Exercise 3.

   (a) What is the "best" hyperparameter combination according to the *AUC* performance metric?

   (b) Use your answer to part (a) and the `Train_data` tibble to fit the RF classifier and obtain the variable importance plot

   (c) Use the fitted RF classifier and `Test_data` to obtain the confusion matrix and performance measures.

   (d) Obtain the ROC and *AUC* for the fitted RF classifier

5. The `hpc_cv` data frame from the **yardstick** package contains resampling results from 10-fold cross-validation for a 4-class problem. The data includes columns for the true class (`obs`), the class prediction (`pred`), class membership probabilities (in columns VF, F, M, and L), and a column that identifies the folds (`Resample`).

   (a) Extract the `hpc_cv` data frame and convert it to tibble.

   (b) Obtain the four PR curves (associated with the four classes) from resampling results in each fold of `Resample`.

   (c) Obtain the resampling estimates of PR *AUC* from results in each fold of `Resample`. Calculate the mean and standard error from the estimated *AUC* values.

A

Appendix

A.1 Required Libraries

We relied quite a bit on functionality in **tidyverse** and **tidymodels** throughout this book. Several core **R** packages in the first meta-package were used for data wrangling and graphics, and a number of those in the second were used for classifier training and testing. You can find out more information on the core packages by consulting the following websites:[1]

- https://www.tidyverse.org/packages/

- https://www.tidymodels.org/packages/

Other sources of information include the books by Wickham and Grolemund [116] and Mailund [75] for **tidyverse**, and Kuhn and Silge [69] for **tidymodels**.

If you have not already done so, you can install these meta-packages

```
install.packages("tidyverse")  # run this once
install.packages("tidymodels") # ditto
```

and load them as follows

```
library(tidyverse)
library(tidymodels)
```

After doing this you can access functionality in the core packages like **dplyr** and **yardstick** without the need to load them individually.[2]

Some of the other required packages for the code examples in this book include the following:

- **magrittr** for data wrangling,

---

[1]Note that the included core packages are listed in the control panel when you load the meta-packages.

[2]Commands for loading of individual core packages were given in some of the code segments in this book. This was done to highlight the relevant package. You can ignore the commands if you have loaded the corresponding meta-package.

- **rpart**, **nnet** and **randomForest** for fitting classifiers,

- **rpart.plot** and **NeuralNetTools** for displaying fitted classifiers

- **ROCR** and **pROC** for ROC analysis,

- **vip** for assessing variable importance.

A.2   Alternative yardstick Functions

As noted in Chapter 2, **yardstick** has two functions that you can use to obtain threshold, overall and composite measures. In that chapter, we focused on functions that return single numerical values for the measures. The use of alternative functions that return a tibble containing the required measure is given below (output omitted).

```
# Using Alternative yardstick Functions

# Class-Specific Measures
DT_pv %>% sens(Survived, pred_class)
DT_pv %>% spec(Survived, pred_class)
DT_pv %>% ppv(Survived, pred_class)
DT_pv %>% npv(Survived, pred_class)

# Overall Measures
DT_pv %>% accuracy(Survived, pred_class)
DT_pv %>% kap(Survived, pred_class)
DT_pv %>% mcc(Survived, pred_class)

# Composite Measures
DT_pv %>% bal_accuracy(Survived, pred_class)
DT_pv %>% j_index(Survived, pred_class)
DT_pv %>% f_meas(Survived, pred_class)
```

A.3   Some Additional R Functions

A.3.1   The con_mat() Function

The following function allows you to create a confusion matrix from a vector of key counts; see (1.9) in Chapter 1 for the definition of this vector for the array summary in Table 1.1. The resulting confusion matrix may be an object of class `"matrix"`, `"table"` or `"conf_mat"` (the default). The function may also be applied to obtain a k-class confusion matrix like what is given in Table 1.6.

```r
con_mat <- function(counts, outcomes, type = "conf_mat") {
  #
  # Create a Predicted by Actual Confusion Matrix
  #
  # Input: counts = vector of cell counts (listed column-wise)
  #        outcome = vector of outcomes
  #        type = "matrix", "table" or "conf_mat"
  #
  # Output: cm = a confusion matrix of class "matrix", "table "
  #              or "conf_mat" (by default)

  if (type == "matrix") {

    k <- sqrt(length(counts))
    cm <- matrix(counts, k, k)
    rownames(cm) <- colnames(cm) <- outcomes

  } else if (type == "table") {

    k <- length(outcomes)
    cm <- counts
    dim(cm) <- c(k,k)
    class(cm) <- "table"
    dimnames(cm) <- list(Predicted=outcomes, Actual=outcomes)

  } else {

    cm <-
      con_mat(counts, outcomes, type = "table") %>%
      vtree::crosstabToCases() %>%
      yardstick::conf_mat(Actual, Predicted,
        dnn = c("Predicted", "Actual"))
  }

  return(cm)
}
```

In the code below, a binary confusion matrix of class `"table"` is created from a 1×4 key count vector, assuming the key counts are in the order given in (1.9). Given the class of the array summary, you can use the row/column profiles approach to extract the key class-specific performance measures from it by using the `prop.table()` function.[3]

```
kcv1 <- c(421, 119, 64, 96) # key count vector
kcv1 %>% con_mat(c("Yes","No"), type = "table")
##          Actual
## Predicted Yes  No
##       Yes 421  64
##        No 119 396
```

Alternatively, you can extract all the performance measures from a given confusion matrix of class `"conf_mat"` as shown below for a 3-class array summary. This is useful when you want to extract performance measures from published confusion matrices in the literature.

```
kcv2 <- c(23,0,0,0,11,0,0,1,9) # key count vector
kcv2 %>% con_mat(LETTERS[1:3]) %T>% print() %>% summary()
##          Actual
## Predicted  A  B  C
##         A 23  0  0
##         B  0 11  1
##         C  0  0  9
## # A tibble: 13 x 3
##    .metric              .estimator .estimate
##    <chr>                <chr>          <dbl>
##  1 accuracy             multiclass     0.977
##  2 kap                  multiclass     0.963
##  3 sens                 macro          0.967
##  4 spec                 macro          0.990
##  5 ppv                  macro          0.972
##  6 npv                  macro          0.990
##  7 mcc                  multiclass     0.964
##  8 j_index              macro          0.957
##  9 bal_accuracy         macro          0.978
## 10 detection_prevalence macro          0.333
## 11 precision            macro          0.972
## 12 recall               macro          0.967
## 13 f_meas               macro          0.968
```

A.3.2   The `kcv_fn()` Function

Given a $k \times k$ confusion matrix with $k > 2$, you can use the following function to obtain the key count vectors for the OvR collection of binary confusion matrices; see Section 1.4.2 for an example of such a collection.

---

[3]Note that the entries of the key count vectors from confusion matrices in Table 1.1 and Table 1.6 are listed column-wise.

```
kcv_fn <- function(cm, i) {
  #
  # Obtain Key Count Vectors for OvR Collection
  #
  # Input: cm = k x k Confusion Matrix as "table" Object
  #        i = 1, 2, ..., k
  #
  # Output: List of Key Count Vectors

  tp <- cm[i, i]; fn <- sum(cm[, i]) - tp
  fp <- sum(cm[i,]) - tp; tn <- sum(cm) - tp - fp - fn

  return(c(tp, fn, fp, tn))
}
```

You can obtain the key count vectors for the OvR collection from Table 1.7 as shown below. The resulting vectors are those for the OvR collection of confusion matrices in Figure 1.6.

```
CM <- c(74, 1, 0, 4, 143, 3, 0, 8, 67) %>%
  con_mat(LETTERS[1:3], type = "table")
1:nrow(CM) %>% map(~ kcv_fn(CM, .))
## [[1]]
## [1]  74   1   4 221
##
## [[2]]
## [1] 143   7   9 141
##
## [[3]]
## [1]  67   8   3 222
```

You can use the following pipeline to obtain the OvR key count vectors, confusion matrices and corresponding row profiles (which yields the OvR *sensitivity* and *specificity* values that you can use to obtain corresponding macro and macro-weighted measures).

```
1:nrow(CM) %>%
  map(~ kcv_fn(CM, .)) %T>%
  print() %>%
  map(~ con_mat(., c("Yes", "No"), type = "table")) %T>%
  print() %>%
  map(~ prop.table(., 2))
```

A.3.3 The acc_ci() Function

The following function allows you to obtain a $100(1-\alpha)\%$ confindence interval for the *Accuracy* performance parameter in Table 2.3; see Section 2.5.1 for the relevant formula. The function allows you to obtain two-sided and one-sided lower confidence intervals.

```
acc_ci <- function(cm, alpha = 0.05, type = "two-sided") {
  #
  # Compute Confidence Interval for Accuracy
  #
  # Input: cm = Confusion Matrix as "table" Object
  #
  # Output: Point Estimate & Confidence Limits

  n <- cm %>% sum()
  x0 <- cm %>% diag() %>% sum()

  estimate = x0 / n
  if (type == "lower") {
    lower_limit = qbeta(alpha, x0, n - x0 + 1)
    upper_limit = 1
  } else {
    lower_limit = qbeta(alpha/2, x0, n - x0 + 1)
    upper_limit = qbeta(1 - alpha/2, x0 + 1, n - x0)
  }

  return(tibble(estimate, lower_limit, upper_limit))
}
```

Given information in the CM confusion matrix below, you can use the following code to obtain 95% two-sided and lower one-sided confidence intervals for the *Accuracy* performance parameter.

```
CM <- con_mat(c(421,119,64,396), c("Yes","No"), type="table")

CM %T>% print() %>% acc_ci() # two-sided interval
##          Actual
## Predicted Yes  No
##       Yes 421  64
##        No 119 396
## # A tibble: 1 x 3
##   estimate lower_limit upper_limit
##      <dbl>       <dbl>       <dbl>
## 1    0.817       0.792       0.841
```

```
CM %>% acc_ci(type = "lower") # lower one-sided interval
## # A tibble: 1 x 3
##   estimate lower_limit upper_limit
##      <dbl>       <dbl>       <dbl>
## 1    0.817       0.796           1
```

A.4 Some Useful Web Links

1. e-Book and Website for **tidyverse**

- **R** for Data Science
  https://r4ds.had.co.nz/

- tidyverse Website
  https://www.tidyverse.org/

2. e-Book and Website for **tidymodels**

- Tidy Modeling with **R**
  https://www.tmwr.org/

- tidymodels Website
  https://www.tidymodels.org/

3. Other e-Books and Websites

- Hands-On Machine Learning with **R**
  https://bradleyboehmke.github.io/HOML/

- ggplot2: Elegant Graphics for Data Analysis (3rd Ed)
  https://ggplot2-book.org/

4. StatQuest Videos by Josh Starmer on Classifiers

- Logistic Regression
  https://www.youtube.com/watch?v=yIYKR4sgzI8

- Decision Trees
  https://www.youtube.com/watch?v=7VeUPuFGJHk

- Decision Trees, Part 2 - Feature Selection and Missing Data
  https://www.youtube.com/watch?v=wpNl-JwwplA

- Neural Networks Pt. 1: Inside the Black Box
  https://www.youtube.com/watch?v=CqOfi41LfDw

- Neural Networks Pt. 2: Backpropagation Main Ideas
  https://www.youtube.com/watch?v=IN2XmBhILt4

- Random Forests in R
  https://www.youtube.com/watch?v=6EXPYzbfLCE

Bibliography

[1] A. Akella and S. Akella. Machine learning algorithms for predicting coronary artery disease: efforts toward an open source solution. *Future Science*, 7, March 2021.

[2] C. Anagnostopoulos, D. J. Hand, and N. M. Adams. Measuring classification performance: the **hmeasure** package. Working Paper, Department of Mathematics, Imperial College, February 2019. A vignette for the **hmeasure** package.

[3] O. Aydemir. A new performance evaluation metric for classifiers: Polygon Area Metric. *Journal of Classification*, 38:16–26, 2021.

[4] G. E. Batista, R. C. Prati, and M.C. Monard. A study of the behaviour of several methods for balancing machine learning training data. *SIGKDD Explorations*, 6:20–29, June 2004.

[5] B. S. Baumer, D. T. Kaplan, and N. J. Horton. *Modern Data Science with* **R**. CRC Press, Taylor & Francis Group, LLC, 2017.

[6] B. Boehmke and B. Greenwell. *Hands-on Machine Learning with* **R**. CRC Press, Taylor & Francis Group, LLC., 2020.

[7] A. P. Bradley. The use of the area under the ROC curve in the evaluation of machine learning algorithms. *Pattern Recognition*, 30(6):1145–1159, 1997.

[8] P. Branco, L. Torgo, and R. P. Ribeiro. A survey of predictive modeling under imbalanced distributions. *ACM Computing Surveys*, 45:1–50, May 2015.

[9] L. Breiman, J. H. Friedman, R. A. Olshen, and C. J. Stone. *Classification and Regression Trees*. Wadsworth, Califonia, USA, 1984.

[10] J. Brownlee. *Imbalanced Classificaton with* **Python**. Jason Brownlee eBook, 2020.

[11] A. Buja, W. Stuetzle, and Y. Shen. Loss functions for binary class probability estimation and classification: structure and applications. Technical report, Statistics Department, The Wharton School, University of Pennsylvania, 2005.

Bibliography

[12] A. Charpentier and S. Tufféry. Statistical learning. In A. Charpentier, editor, *Computational Actuarial Science with* **R**, pages 165–205. Taylor & Francis Group, LLC, 2015.

[13] N. V. Chawla, K. W. Bowyer, L. O. Hall, and W. P. Kegelmeyer. SMOTE: Synthetic minority over-sampling technique. *Journal of Artificial Intelligence Research*, 16:321–357, 2002.

[14] D. Chicco and G. Jurman. The advantages of the Matthews correlation coefficient (MCC) over F1 score and accuracy in binary classification evaluation. *BMC Genomics*, 21(6):1–13, 2020.

[15] D. Chicco and G. Jurman. The Matthews correlation coefcient (MCC) should replace the ROC AUC as the standard metric for assessing binary classifcation. *BioData Mining*, 16(4):1–23, 2023.

[16] P. Cichosz. Assessing the quality of classification models: performance measures and evaluation procedures. *Central European Journal of Engineering*, 1(2):132–158, 2011.

[17] D. A. Cieslak and N. V. Chawla. Learning decision trees for unbalanced data. In *Proceedings of the 2008 European Conference on Machine Learning and Knowledge Discovery in Databases-Part I*, pages 241–256, 2008.

[18] D. A. Cieslak, T. R. Hoens, N. V. Chawla, and W. P. Kegelmeyer. Hellinger distance decision trees are robust and skew-insensitive. *Data Mining and Knowledge Discovery*, 24(1):136–158, 2012.

[19] J. Davis and M. Goadrich. The relationship between precision-recall and ROC curves. In W. W. Cohen and A. Moore, editors, *Proceedings of the 23rd International Conference on Machine Learning*, pages 233–240, 2006.

[20] A. C. Davison and D. V. Hinkley. *Bootstrap Methods and their Application*. Cambridge Univesity Press, Oxford, 1997.

[21] R. Delgado and J. D. Núñez González. Enhancing confusion entropy CEN for binary and multiclass classification. *PLoS ONE*, 14(1):1–30, 2019.

[22] R. Delgado and Xavier-Andoni Tibau. Why Cohen's kappa should be avoided as performance measure in classification. *PLoS ONE*, 14(9):1–26, 2019.

[23] T. J. DiCiccio and B. Efron. Bootstrap confidence intervals. *Statistical Science*, 11(3):189–228, 1996.

[24] T. G. Dietterich. Approximate statistical tests for comparing supervised classification learning algorithms. *Neural Computation*, 10:1895–1923, 1998.

[25] H. Dong, A. Supratak, W. Pan, C. Wu, P. M. Matthews, and Y. Guo. Mixed neural network approach for temporal sleep stage classification. *IEEE Transaction on Neural Systems Rehabilitation Engineering*, 26(2):324–333, 2018.

[26] C. Drummond and R.C. Holte. Explicitly representing expected cost: an alternative to ROC representation. In D.E. Losada and J.M. Fernández-Luna, editors, *Proceeding of the Sixth ACM SIGKDD International Conference on Knowledge Discovery and Data Mining*, pages 198–207, 2000.

[27] D. C. Edwards, C. E. Metz, and M. A. Kupinski. Ideal observers and optimal ROC hypersurfaces in N-class classification. *IEEE Transactions on Medical Imaging*, 23:891–895, April 2004.

[28] B. Efron. Bootstrap methods: another look at the jacknife. *The Annals of Statistics*, 7(1):1–26, 1979.

[29] B. Efron. Estimating the error rate of a prediction rule: improvement on cross-validation. *Journal of the American Statistical Association*, 78(382):316–331, June 1983.

[30] B. Efron and R. Tibshirani. Improvements on cross-validation: the .632+ bootstrap method. *Journal of the American Statistical Association*, 92(438):548–560, June 1997.

[31] S. O. Ejeh, O. O. Alabi, O. O. Ogungbola, O. O. Olatunde, and Z. O. Dere. A comparison of multinomial logistic regression and artificial neural network classification techniques applied to TB/HIV data. *American Journal of Epidemiology and Public Health*, 6:12–18, April 2022.

[32] C. Elkan. The foundations of cost-sensitive learning. In *Proceedings of the Seventeenth International Joint Conference on Artificial Intelligence*, pages 973–978, May 2001.

[33] B. Engelmann, E. Hayden, and D. Tasche. Measuring the discriminative power of rating systems. In *Social Science Research Network*, 2003.

[34] T. Fawcett. ROC graphs: notes and practical considerations for data mining researchers. Technical Report HPL-2003-4, HP Laboratories Tech. Rep., Palo Alto, California, 2003.

[35] T. Fawcett. An introduction to ROC analysis. *Pattern Recognition Letters*, 27:861–874, 2006.

[36] A. Fernández, S. García, M. Galar, R. C. Prati, B. Krawczyk, and F. Herrera. *Learning from Imbalanced Data Sets*. Springer Nature Switzerland AG, 2018.

[37] C. Ferri, J. Hernández-Orallo, and R. Modroiu. An experimental comparison of performance measures for classification. *Pattern Recognition Letters*, 30(1):27–38, January 2009.

[38] C. Ferri, J. Hernández-Orallo, and M. A. Salido. Volume under the ROC surface for multi-class problems. Exact computation and evaluation of approximations. In *Proceedings of 14th European Conference on Machine Learning*, pages 108–120, 2003.

[39] R. A. Fisher. The use of multiple measurements in taxonomic problems. *Annals of Eugenics*, 7:179–188, 1936.

[40] P. A. Flach. The geometry of ROC space: understanding machine learning metrics through ROC isometrics. In *Proceedings of the International Conference on Machine Learning (ICML)*, pages 194–201, January 2003.

[41] J. Fox and S. Weisberg. *An **R** Companion to Applied Regression*. SAGE Publications, Inc., 2011.

[42] K. Fukunaga. *Introduction to Statistical Pattern Recognition*. Elsevier, Inc., second edition, 1990.

[43] M. Grandini, E. Bagli, and G. Visani. Metrics for multi-class classification: an overview. White Paper, Dipartimento di Ingegneria e Scienze Informatiche, Universita degli Studi di Bologna, Italy, 2020.

[44] P. Gupta and D. D. Seth. Improving the prediction of heart disease using ensemble learning and feature selection. *International Journal of Advances in Soft Computing and its Applications*, 14(2):36–48, July 2022.

[45] C. Halimu, A. Kasem, and S. H. Shah Newaz. Empirical comparison of area under ROC curve (AUC) and Matthews correlation coefficient (MCC) for evaluating machine learning algorithms on imbalanced datasets for binary classification. In *Proceedings of the 3rd International Conference on Machine Learning and Soft Computing*, pages 1–6, 2019.

[46] D. Hand and P. Christen. A note on using the F-measure for evaluating record linkage algorithm. *Statistics and Computing*, 28(3):539–547, 2018.

[47] D. J. Hand. *Discrimination and Classification*. John Wiley, Chichester, 1981.

[48] D. J. Hand. *Construction and Assessment of Classification Rules*. John Wiley & Sons Ltd, Chichester, 1997.

[49] D. J. Hand. Classifier technology and the illusion of progress. *Statistical Science*, 21(1):1–15, 2006.

[50] D. J. Hand. Measuring classifier performance: a coherent alternative to the area under the ROC curve. *Machine Learning*, 77(1):103–123, 2009.

[51] D. J. Hand. Evaluating diagnostic tests: the area under the ROC curve and the balance of errors. *Statistics in Medicine*, 29(1):1502–1510, 2010.

[52] D. J. Hand. Assessing the performance of classification methods. *International Statistical Review*, 80:400–414, 2012.

[53] D. J. Hand and C. Anagnostopoulo. Notes on the H-measure of classifier performance. *Advances in Data Analysis and Classification*, 17:109–124, April 2023.

[54] D. J. Hand and C. Anagnostopoulos. A better beta for the H-Measure of classification performance. *Pattern Recognition Letters*, 40:41–46, April 2014.

[55] D. J. Hand and W. E. Henley. Statistical classification methods in consumer credit scoring: a review. *Journal of the Royal Statistical Society A*, 160:523–541, 1997.

[56] D. J. Hand and R. J. Till. A simple generalisation of the area under the ROC curve for multiple class classification problems. *Machine Learning*, 45:171–186, 2001.

[57] J. A. Hanley and B. J. McNeil. The meaning and use of the area under a receiver operating characteristic ROC curve. *Radiology*, 143(1):29–36, April 1982.

[58] P. Hart. The condensed nearest neighbor rule. *IEEE Transactions on Information Theory*, 14(3):515–516, 1968.

[59] T. Hastie, R. Tibshirani, and J. Friedman. *The Elements of Statistical Learning Data Mining, Inference, and Prediction.* Springer Science + Business Median, New York, second edition, 2017.

[60] J. Hilden. The area under the ROC curve and its competitors. *Medical Decision Making*, 11:95–101, June 1991.

[61] J. Huang and C. X. Ling. Using AUC and accuracy in evaluating learning algorithms. *IEEE Transactions on Knowledge and Data Engineering*, 17(3):299–310, March 2005.

[62] G. James, D. Witten, T. Hastie, and R. Tibshirani. *An Introduction to Statistical Learning with Applications in* **R**. Springer Science + Business Median New York, second edition, 2021.

[63] Y. Jiao and P. Du. Performance measures in evaluating machine learning based bioinformatics predictors for classification. *Quantitative Biology*, 4:320–330, 2016.

[64] G. Jurman, S. Riccadonna, and C. Furlanello. A comparison of MCC and CEN error measures in multi-class prediction. *PLoS ONE*, 7(8):1–8, 2012.

[65] J. Kapasný and M. Řezáč. Three-way ROC analysis using SAS software. *Acata Universitatis Agriculturae et silviculturae Mendelianae Brunensis*, LXI(7):2269–2275, 2013.

[66] R. S. Kleiman and D. Page. AUC_μ: A performance metric for multi-class machine learning models. In K. Chaudhuri and R. Salakhutdinov, editors, *Proceedings of the 36-th International Conference on Machine Learning*, volume 97, pages 3439–3447. PMLR, June 2019.

[67] R. Kohavi. A study of cross-validation and bootstrap for accuracy estimation and model selection. In *Proceedings of the Fourteenth International Joint Conference on Artificial Intelligence (IJCAI-95)*, volume 2, pages 1137–1143. Morgan Kaufmann Publishers Inc., March 1995.

[68] M. Kuhn and K. Johnson. *Feature Engineering and Selection: A Practical Approach to Predictive Models*. CRC Press, Taylor & Francis Group, LLC, 2020.

[69] M. Kuhn and J. Silge. *Tidy Modeling with **R***: A Framework for Modeling in the Tidyverse*. O'Reilly Media, Inc., 2022.

[70] T. C. W. Landgrebe and R. P. W. Duin. A simplified extension of the area under the ROC to the multiclass domain. In A. Louw, N. Kleynhans, and N. Zulu, editors, *Proceedings 17th Annual Symposium of the Pattern Recognition Association of South Africa*, pages 241–245. The International Association for Pattern Recognition, 2006.

[71] E. L. Lehmann. *Elements of Large-Sample Theory*. Springer-Verlag New York, Inc., 1999.

[72] J. Li and J. P. Fine. ROC analysis with multiple classes and multiple tests: methodology and its application in microarray studies. *Biostatistics*, 9(3):556–576, 2008.

[73] S. Liu, H. Zhui, K. Yi, X. Sun, W. Xu, and C. Wang. Fast and unbiased estimation of volume under ordered three-class ROC surface (VUS) with continuous or discrete measurements. *IEEE Access*, 8:136206–136222, 2020.

[74] Y. Liu, Y. Zhou, S. Wen, and C. Tang. A strategy on selecting performance metrics for classifier evaluation. *International Journal of Mobile Computing and Multimedia Communications*, 6:20–35, 2014.

[75] T. Mailund. **R** *Data Science Quick Reference: A Pocket Guide to APIs, Libraries, and Packages*. Apress, 2019.

[76] J. N. Mandrekar. Receiver operating characteristic curve in diagnostic test assessment. *Journal of Thoracic Oncology*, 5(9):1315–1316, September 2010.

[77] I. Markoulidakis, I. Rallis, I. Georgoulas, G. Kopsiaftis, A. Doulamis, and N. Doulamis. Multiclass confusion matrix reduction method and its application on net promoter score classification problem. *Technologies*, 9(4):1–22, November 2021.

[78] B. W. Matthews. Comparison of the predicted and observed secondary structure of T4 phage lysozyme. *Biochimica Biophysica Acta*, 405(2):442–451, 1975.

[79] I. McNemar. Note on the sampling error of the difference between correlated proportions or percentages. *Psychometrika*, 12:153–157, 1947.

[80] I. D. Mienye, Y. Sun, and Z. Wang. An improved ensemble learning approach for the prediction of heart disease risk. *Informatics in Medicine Unlocked*, 20:1–5, 2020.

[81] A. M. Mood, F. A. Graybill, and D. C. Boes. *Introduction to the Theory of Statistics*. McGraw-Hill, Inc., third edition, 1974.

[82] F. S. Nahm. Receiver operating characteristic curve: overview and practical use for clinicians. *Korean Journal of Anesthesiology*, 75(1):25–36, 2022.

[83] F. Nwanganga and M. Chapple. *Practical Machine Learning in R*. John Wiley & Sons, Inc., Indianapolis, Indiana, 2020.

[84] N. A. Obuchowski. Receiver operating characteristic curves and their use in radiology. *Radiology*, 229:3–8, 2003.

[85] N. A. Obuchowski, M. L. Lieber, and F. W. Wians, Jr. ROC curves in clinical chemistry: uses, misuses, and possible solutions. *Clinical Chemistry*, 50(7):1118–1125, 2004.

[86] I. Olkin, L. J. Gleser, and C. Derman. *Probability Models and Applications*. Macmillan College Publishing Company, Inc., second edition, 1994.

[87] S. Pintea and R. Moldovan. The receiver-operating characteristic (ROC) analysis: fundamentals and applications in clinical psychology. *Journal of Cognitive and Behavioral Psychotherapies*, 9(1):49–66, March 2009.

[88] D. M. W. Powers. Evaluation: from precision, recall and F-factor to ROC, informedness, markedness & correlation. Technical Report SIE-07-001, School of Informatics and Engineering, Flinders University, Adelaide, Australia, December 2007.

[89] A. D. Pozzolo, R. Johnson, O. Caelen, S. Waterschoot, N. V. Chawla, and G. Bontempi. Using HDDT to avoid instances propagation in unbalanced and evolving data streams. In *Proceedings of the International Joint Conference on Neural Networks*, pages 588–594, July 2014.

[90] F. Provost and T. Fawcett. *Data Science for Busines: What You Need to Know About Data Mining and Data-Analytic Thinking*. O'Reilly Media, Inc., 2013.

[91] F. Provost, T. Fawcett, and R. Kohavi. The case against accuracy estimation for comparing induction algorithms. In J. Shavlik, editor, *Proceedings of the Fifteenth International Conference on Machine Learning*, pages 445–453, 1998.

[92] J. R. Quinlan. *C4.5: Programs for Machine Learning*. Morgan Kayfmann, San Mateo, CA, 1993.

[93] V. Raghavan, G. S. Jung, and P. Bollmann. A critical investigation of recall and precision as measures of retrieval system performance. *ACM Transactions on Information Systems*, 7(3):205–229, 1989.

[94] G. M. Reaven and R. G. Miller. An attempt to define the nature of chemical diabetes using a multidimensional analysis. *Diabetologia*, 16:17–24, 1979.

[95] H. I. Rhys. *Machine Learning with* **R**, *the tidyverse and mlr*. Manning Publications Co., 2020.

[96] M. L. Rizzo. *Statistical Computing with* **R**. Chapman & Hall/CRC, second edition, 2019.

[97] R. J. Roiger and M. W. Geatz. *Data Mining: A Tutorial-Based Primer*. Pearson Education, Inc., 2003.

[98] M. S. Roulston. Performance targets and the Brier score. *Meteorological Applications*, 14:185–194, 2007.

[99] T. Saito and M. Rehmsmeier. The Precision-Recall plot is more informative than the ROC plot when evaluating binary classifiers on imbalanced datasets. *PLoS ONE*, 10(3):1–21, 2015.

[100] B. K. Scurfield. Multiple-event forced-choice tasks in the theory of signal detectability. *Journal Mathematical Psychology*, 40(3):253–269, 1996.

[101] T. Sing, O. Sander, N. Beerenwinkel, T. Lengauer, T. Unterthiner, and F. G. M. Ernst. Package ROCR, October 2022.

[102] M. Sokolova, N. Japkowicz, and S. Szpakowicz. Beyond accuracy, F-score and ROC: a family of discriminant measures for performance evaluation. In *Proceedings of 19th the ACS Australian Joint Conference on Artificial Intelligence*, pages 1015–1021, 2006.

[103] M. Sokolova and G. Lapalme. A systematic analysis of performance measures for classification tasks. *Information Processing and Management*, 45:427–437, 2009.

[104] P. Sonego, A. Kocsor, and S. Pongor. ROC analysis: applications to the classification of biological sequences and 3D structures. *Briefings in Bioinformatics*, 9(3):198–209, 2008.

[105] R. M. Stein. The relationship between default prediction and lending profits: integrating ROC analysis and loan pricing. *Journal of Banking and Finance*, 29:1213–1236, 2005.

[106] K. Takahashi, K. Yamamoto, A. Kuchiba, and T. Koyama. Confidence interval for micro-averaged F1 and macro-averaged F1-scores. *Applied Intelligence*, 52(1):4961–4972, 2022.

[107] P. N. Tattar. *Hands-On Ensemble Learning with **R***: *A Beginner's Guide to Combining the Power of Machine Learning Algorithms Using Ensemble Techniques*. Packt Publishing, 2018.

[108] A. Tharwat. Classification assessment methods. *Applied Computing and Informatics*, 17(1):168–192, 2021.

[109] I. Tomek. Two modifications of CNN. *IEEE Transactions on Systems, Man, and Communications*, 7(2):679–772, November 1976.

[110] G. Vanwinckelen and H. Blockeel. On estimating model accuracy with repeated cross-validation. In *BeneLearn 2012: Proceedings of the 21st Belgian-Dutch Conference on Machine Learning*, pages 39–44, 2012.

[111] A. Vellido, P. J. G. Lisboa, and J. Vaughan. Neural networks in business: a survey of applications (1992-1998). *Expert Systems with Applications*, 17(1):51–70, 1999.

[112] M. S. Wandishin and S. J. Mullen. Multiclass ROC analysis. *Weather and Forecasting*, 24:530–547, 2009.

[113] A. R. Webb and K. D. Copsey. *Statistical Pattern Recognition*. John Wiley & Sons, Ltd, third edition, 2011.

[114] J. M. Wei, X. J. Yuan, Q. H. Hu, and S. Q. Wang. A novel measure for evaluating classifiers. *Expert Systems with Applications*, 37:3799–3809, 2010.

[115] G. M. Weiss and F. Provost. Learning when training data are costly: the effect of class distribution on tree induction. *Journal of Artificial Intelligence Research*, 19:315–354, 2003.

[116] H. Wickham and G. Grolemund. **R** *for Data Science: Import, Tidy, Transform, Visualize, and Model Data*. O'Reilly Media Inc., 2017.

[117] B. Widrow, D. E. Rumelhart, and M. A. Lehr. Neural networks: applications in industry, business and science. *Communications of the ACM*, 37(3):93–105, 1994.

[118] D. Zhang, J. Wang, and X. Zhao. Estimating the uncertainty of average F1 scores. In *Proceedings of the 2015 International Conference on the Theory of Information Retrieval*, pages 317–320, 2015.

[119] N. Zumel and J. Mount. *Practical Data Science with* **R**. Manning Publications Co., second edition, 2020.

Index

Printed in the United States
by Baker & Taylor Publisher Services